Sugarcane Ratoon Management

Dr. R. S. Verma
Principal Scientist (Agronomy)
Indian Institute of Sugarcane Research
Lucknow
and
Ex. Director
Directorate of Sugarcane Development
Ministry of Agriculture, Government of India

International Book Distributing Co.
Chaman Studio Building, 2nd Floor, Charbagh
Lucknow-226004, U.P. (India)

Published by
INTERNATIONAL BOOK DISTRIBUTING COMPANY
(Publishing Division)
Chaman Studio Building, 2nd Floor,
Charbagh, Lucknow - 226 004 (India)
Tel. : 450004, 450007, 459058 Fax : 0522-458629
E-Mail : ibdco@sancharnet.in

First Edition 2002

ISBN 81-85860-83-1

Composed and Designed at :
Panacea Computers
33, Nehru Road, Sadar Cantt.
Lucknow - 226 002 U.P.
Tel. : 0522-481164, 480546
E-mail : prasgupt@rediffmail.com

Printed at :
Army Printing Press
33, Nehru Road, Sadar Cantt
Lucknow - 226 002
Tel. : 0522- 481164, 480546
E-mail : armyprintingpress@rediffmail.com

Dedicated to
the respectful memory of
my mother
"Ammaji"

Acknowledgements

Data are the source material from which all scientific concepts arise. Thus I wish to acknowledge all the researchers whose data have been used in preparing this book as it is a product of the work of many dedicated people.

I also express my thanks to Mr. G.D. Dhariyal and Mr. G. K.Gupta for providing help in searching the scattered published resource material in the library and to Mr. Vipin Dhawan for photographic work

Author

Contents

List of figures

1. Ratooning in General

1.1. Historical background

Ratoon cropping is an age old system which has been in practice for many years. Although the origin of ratooning is probably not known for any particular crop, it may have begun when man first noticed regrowth of new shoots following cutting of certain crops at harvest, thus producing a new crop without replanting. Also, early man's observations of grass-land regrowth following burning might have created an interest in utilizing regrowth of plants as a basis of multiple harvest from a original root system (Plucknett *et al*., 1970). Ratooning of sugarcane has been in vogue in almost all of the cane growing countries of the world, except Java (now Indonesia). In India, practice of ratooning has been known from the very early times. However, the area under ratoon was not as high as it is today (Tandon and Misra, 1954). According to the Indian Sugar Committee Report, 1920, ratooning of cane was not a very common practice in our country. It was due to the fact that the cane varieties under cultivation at that time were wholly indigenous and were very poor ratooners (Rattan and Singh, 1954). The introduction of Co canes, which are known to be better ratooner, revolutionised the position and lead to the rapid increase in the area under ratoons (Prasada Rao,1980). In Uttar Pradesh, an important sugarcane growing state in subtropics, acreage occupied by the ratoons increased from 33.5 per cent in 1955-56 to 55.3 per cent in 1992-93. The ratoon productivity which was 66 per cent of plant crop in 1955-56 increased to 97 per cent in 1992-93.

1.2. **Nomenclature**

The word ratoon seems to have originated either from the Latin words *retonsus* – cut down or mown (Lewis and Short, 1958); or *retono* – to thunder back, resound (Anonymous, 1950); the Spanish *retono* — a fresh shoot or sprout (Anonymous, 1960); or even the French *rejeton* – a sucker or shoot, scion, descendant, off-spring, or sprout (Plucknett *et al.*, 1970).

1.3. **Definition**

The Oxford Dictionary defines ratoon as " a new shoot or sprout springing up from the root of sugarcane after it has been cropped". This definition needs correction as regrowth does not occur from the roots of the plant, but rather from the buds on the underground portion of the stem. Winburne (1962) defined the term as "a basal sucker for propagation as in sugarcane, pineapple and banana". Encyclopaedia Britannica (Macropaedia, 17 : 769) described ratooning in sugarcane as "when the cane is harvested a portion of stalk is left underground to give rise to a succeeding growth of cane, the ratoon or stubble crop".

Basically ratoon cropping implies :

i) more than one harvest from a single planting,

ii) regrowth from basal buds on the stem, and

iii) harvesting of most or all the aerial portion of the plant.

The first harvest of a crop is usually called the "plant crop" and each succeeding harvest is designated "first ratoon", "second ratoon" etc.

1.4. **Advantages**

The advantage of ratooning over the plant crop in

sugarcane lies in its reduced cost of production, high sugar recovery during early periods of crushing and reduced crop life cycle. Both the growers and the millers gain through the practice of ratooniong, the former get larger benefit because of reduction in cost of cultivation and the latter due to better recovery per cent cane. In *Gur* tract it is the ratoon crop which is crushed early in October-November to take advantage of high prices of *Gur* in the market.

(a) **Reduced production cost :** Ratoon plays a very important role in economising the cost of sugarcane production, through saving in the cost of preparatory cultivation, seed material and planting expenses. The savings range from 25 to 30 per cent of the total cost of cultivation. At Sugarcane Research Station, Anakapalle (Andhra Pradesh) the cultivation cost of first ratoon of sugarcane (Cv Co 419) worked out by Lakshmikantham(1965) was 76.6 per cent of the plant crop.

(b) **High sugar recovery :** Ratoons mature 3-4 week earlier than plant cane and give better juice quality and sugar recovery during early crushing period in comparison to plant crop of the same variety grown under similar conditions (Vasudeva Rao *et al.*, 1946; Shrikhande *et al.*, 1958). Thus the ratoons play the role of early varieties. Dutt (1946) estimated that in tracts where over 30 per cent of the cane crop crushed is from ratoon, the recovery is increased by about 1 per cent, provided it is crushed early in the season.

c) **Early vacation of field :** Early maturity due to reduced crop life enables the farmer to harvest the ratoon crop early in the season. Early harvesting provides an opportunity to raise another crop in the same field to increase cropping intensity for efficient land use. In subtropical India, for example, a good crop of wheat is grown in the fields of November-December harvested ratoons.

1.5. Disadvantages

The Disadvantages associated with ratooning are:

(a) **Shallow root system :** Ratoons are reported to have shallower root system than plant cane (Fig. 1.1). Due to shallow root system ratoons are less resistant to drought (Dillewijin, 1953; Prasada Rao, 1980) and more prone to lodging with less nutrient foraging capacity than plant crop.

(b) **More fibre content :** Ratoons are noted to have more fibre (Table 1.1) than plant crop (Gupta, 1981). Mohan Rao *et al.*, 1956 at Anakapalle (Andhra Pradesh) also obtained 1.15 and 0.69 per cent higher fibre content in ratoon than plant canes during November and December, respectively, in variety Co 419. Higher fibre content in canes lowers the juice extraction per cent.

Fig. 1.1. Root systems of plant and ratoon crops of sugarcane

Table 1.1. Fibre per cent in cane stalks of plant crop and ratoon (Cv Co 1148)

Crop	December	February	April
Plant Crop	10.96	12.24	13.41
Ratoon	11.14	13.26	15.81

Source :Gupta, 1981

(c) **Early stoppage of growth :** It is a commonly known fact that under indentical conditions and treatments if the ratoons and plant crops of the same variety are allowed to grow from the beginning, the former complete their different phases of the life cycle earlier. Because of this ratoons are bound to yield less.

(d) **Lesser nitrogen use efficiency :** Ratoons are less efficient utilizer of applied nitrogen than plant crop. It has been observed that for the induction of the same amount of nitrate reductase activity *in vivo*, a ratoon crop requires relatively more NO_{3-} nitrogen (Rai *et al.*, 1989) than the corresponding plant crop.

(e) **More arrowing and pith formation :** Arrowing which causes cessation of growth commences earlier in ratoons than plant crop and the extent is also more (Naidu and Lakshmikantham, 1946). Ratoons were also found to contain more pith than plant crop.

(f) **Build up of more insect pests and diseases :** Ratoons are known to build up more pests and diseases than plant crop. This is apparently due to facilities for continuation of pests or diseases in an uninterrupted manner because of availability of host throughout for feeding at the same place (Prasada Rao, 1980). Incidence of insect pests like top borer, white fly and stem borer was more in ratoon than in plant cane (Mukherjee, 1952). Of late, lygaid bugs have been seen playing havoc in ratoons during hot weather in western Uttar Pradesh (Tandon *et al.*, 1957). The build up of diseases like smut, grassy shoot, ratoon stunting and also red rot is cumulative and appear fast in ratoon. With the increase in number of ratoons the incidence of these diseases were more (Srivastava and Seth, 1990). Smut is yet another problem in ratooning. While the per cent reduction in cane yields due to smut infection was found to be 49 per cent in plant cane, it increased to 62 per cent in succeeding ratoon (Prasada Rao, 1980). Red rot affected stubbles of the previous season remaining in the field were held responsible for the spread of the diseases in the succeeding season. Thus if the crop is affected by insects or by some diseases like red rot, smut, ratoon stunting and grassy shoot, it is better to plough out and clear the whole field and the affected crop should not be ratooned.

1.6. **Area and productivity**

Ratoons account for sizeable share in the total area of sugarcane in India. The data available separately regarding area, production and productivity of ratoon and plant crops for Uttar Pradesh and Punjab (Subtropical States) and Tamil Nadu, Karnataka and Orissa (Tropical States) are presented in tables 1.2 and 1.3. In general, proportion of area under ratoons is higher in the subtropical states compared to tropical ones. In Uttar Pradesh and Punjab, about 55 per cent of the total cane area is presently under ratoons whereas due to better emphasis on production per hectare, area under ratoon is only 35 to 40 per cent in Tamil Nadu, Karnataka and Orissa which is considerably lower than Uttar Pradesh and Punjab. According to an estimate, about 50 per cent of the cane area in the country is under ratoons (Rao, 1981). However, the productivity of ratoon is lower as compared to plant canes.

Table 1.2. Relative area and productivity of plant and ratoon crops in the reserved areas of Uttar Pradesh (Subtropical India)

Year	Sugarcane acreage (lac ha)	Per cent area under ratoon crop	Ratoon yield as percentage of plant crop
1955-56	7.00	33.5	66.0
1956-57	-	-	70.1
1957-58	-	-	68.9
1958-59	-	-	67.4
1959-60	-	-	70.5
1960-61	9.28	37.3	64.7
1961-62	9.84	39.5	75.4
1962-63	8.64	45.5	81.5
1963-64	8.12	37.0	73.2
1964-65	9.83	37.5	74.7
1965-66	10.31	40.6	71.0
1966-67	8.26	43.2	73.3
1967-68	6.56	39.0	75.6
1968-69	8.79	35.3	76.0
1969-70	10.46	40.8	75.2
1970-71	10.78	47.6	78.5
1971-72	9.55	48.8	82.0
1972-73	10.48	48.5	78.0
1973-74	12.08	44.5	86.2
1974-75	12.53	50.7	77.1
1975-76	12.15	48.9	88.6
1976-77	11.62	48.4	82.0
1977-78	13.79	47.7	90.1
1978-79	14.67	57.2	91.6

1979-80	12.33	61.4	83.5
1980-81	12.05	52.2	82.8
1981-82	5.39	48.6	86.9
1982-83	17.32	55.7	92.1
1983-84	16.40	58.5	93.0
1984-85	14.70	57.5	96.2
1985-86	14.34	53.6	86.8
1986-87	16.59	50.3	92.5
1987-88	18.50	54.8	92.6
1988-89	18.48	56.4	93.3
1989-90	18.34	55.0	97.9
1990-91	18.30	52.5	99.6
1991-92	18.55	52.5	96.9
1992-93	18.07	55.3	97.0

(Source: Personal communication from the Office of Cane Commissioner, Uttar Pradesh, India)

Table 1.3: Relative area and productivity of plant and ratoon crops in Punjab, Tamil Nadu, Karnataka and Orissa

Year	Sugarcane area (lac ha)	Per cent area under ratoon crop	Ratoon yield as percentage of plant crop
		Punjab	
1988-89	0.98	55.1	-
1989-90	1.03	56.3	-
1990-91	1.01	59.4	-
1991-92	1.09	45.9	-
1992-93	1.35	54.8	-
		Tamil Nadu	
1988-89	2.21	35.5	86.5
1989-90	2.32	35.3	81.2

1990-91	2.39	33.5	80.4
1991-92	2.32	33.9	86.1
1992-93	1.61	43.5	-
		Karnataka	
1988-89	2.39	42.7	89.5
1989-90	2.59	41.3	94.7
1990-91	2.68	40.3	98.8
1991-92	2.24	39.7	100.0
1992-93	1.90	40.0	100.0
		Orissa	
1988-89	0.47	38.3	55.7
1989-90	0.48	38.9	59.8
1990-91	0.49	38.8	63.7
1991-92	0.51	37.2	64.5
1992-93	0.52	+36.5	64.4

(Source: Sugarcane Statistics in India, Published by Directorate of Sugarcane Development, Ministry of Agriculture, Govt. of India, 1993 pp. 68-71)

1.7. Causes responsible for low ratoon yields

The cane yields from ratoon crop are noted to be, in general, less than that of plant crop. Factors responsible for poor yield of the ratoons are generally :

(a) **Neglected attitude of the growers :** Neglected attitude of the growers is to a large extent responsible for poor performance of ratoon crops (Ramakrishna Rao, 1960; Adhate, 1960; Verma, 1981). Some farmers regard the ratoon crop as "God's gift " and do not follow proper cultural, manurial, irrigational and plant protection schedules, with the result many ratoon fields do not yield satisfactorily.

(b) **Piecemeal harvesting :** In reserved areas of the sugar factories, cane is supplied to the mills by the local cane unions of the farmers under the supervision of the state, through a system of bonding and equitable distribution of cutting-permits issued by the cane unions on tonnage. Due to the system of equitable purchase, it is evidently not possible for farmer to harvest his complete field at a time. Consequently, they have to harvest their crop in piecemeal in small patches throughout the crushing season starting from October or early November to April or May. By the time of last harvesting the stubbles of previous harvestings have already sprouted. Therefore, most of these fields have ratoon clumps of different size in patches due to wide gap in time of harvesting. This does not permit desirable crop husbandry for the ratoon because the whole field is never fully free for operations.

(c) **Poor tilth due to compaction :** Poor tilth due to compaction of soil in the root zone of the crop leads to depletion of vigour of root system (Prasada Rao, 1980). The loose and friable soil at the beginning of the plant crop gradually becomes firm and compact at the commencement of ratoon crop because of alternate wetting and drying and also compaction caused during the harvest of the plant crop by labourers. As a result, ratoon experience inadequate soil aeration. This condition decreases intake of water and restricts nutrient uptake by roots. In sugarcane grown on medium to fine textured soils, irrigation should be cut off during harvesting. Withholding water supply decreases the soil moisture content and in this way severe compaction can be avoided.

(d) **Narrow optimum time for ratoon start :** There is specified period (February-March) to start ratoons for obtaining good yields. Ratoons started from November to January in subtropical India, do not give good yields because of excessive gaps due to poor sprouting under low temperature conditions. Similarly, ratoons started in summer months (April to May) also yield low because of shorter pre-*monsoon* period available for formative phase of the crop.

1.8. **Is ratooning a justifiable practice?**

In order to answer the question whether ratooning is a justifiable practice, and if so, for how many years and under what conditions of management, experiments were conducted at Kalai (Aligarh, Uttar Pradesh, India) for 10 years (from1939-49). Results obtained indicated that first year ratoons, if grown with proper management and inputs, yield almost equally as the plant cane. However, the yields of 2nd and 3rd year ratoons were appreciably lower (Mukherjee, 1952). According to Dutt (1946) and Mohan Rao *et al.* (1956), ratooning is a justifiable practice provided sufficient care is taken of the cultural, manurial, insect pests and disease management aspects. It is evidently established through a number of experiments carried over at different locations in the cane growing belts that if ratoon crop is grown adopting recommended production and protection practices, the yields obtained are almost as good as those obtained with plant cane (Kapoor,1966 and Mathur, 1986). This is generally applicable with the ratoons taken from plant canes planted within the normal planting season (February-March) but in the late planting (April-May) which is generally done in western Uttar Pradesh after the harvest of *rabi* crops, ratoons give higher cane yields than late planted canes. Instances of the first and second ratoons giving higher yields than the plant cane are also reported by Kadirvel (1972) in case of varieties Co 419 and Co 449 and Mane and Salunkhe (1978) in case of Co 740.

As regards the multiplication and carry over of insect pests by ratoons, work done at Kalai (Aligarh, Uttar Pradesh) during 1939-40 to 1949-50 showed that borers and pyrilla first bred in ratoons in March-April and then migrated to plant cane in May and June (Tandon *et al.*, 1957). However, segregated plant cane was in no way less affected with these pests than the one grown contiguous to ratoon. It clearly indicates that prohibition of ratoon is not likely to control them. In the absence of ratoon even the unploughed stubbles with their young sprouts and the standing crop for late harvest were found to serve as breeding grounds for these pests (Tandon and Misra 1954). Thus, only ratoons cannot be held responsible of carrying over pests and diseases from season to season.

Regarding diseases it may be stated that these can greatly be kept under control by using disease-free seed material of resistant varieties for planting of plant crop. Thorough rouging of infected stools for plant crop proves helpful in checking their build up in ratoons. Thus ratoons, if taken after healthy plant crop, do not necessarily encourage diseases more than plant crop.

2. Stubble morphology, growth and development of ratoons

2.1. Stubble morphology

Plant parts left in the ground after the harvesting of sugarcane crop constitute sugarcane stubbles (Fig. 2.1). These stubbles consist of underground portion of stalks having closely spaced joints each with its complement of bud (Fig.2.2) and

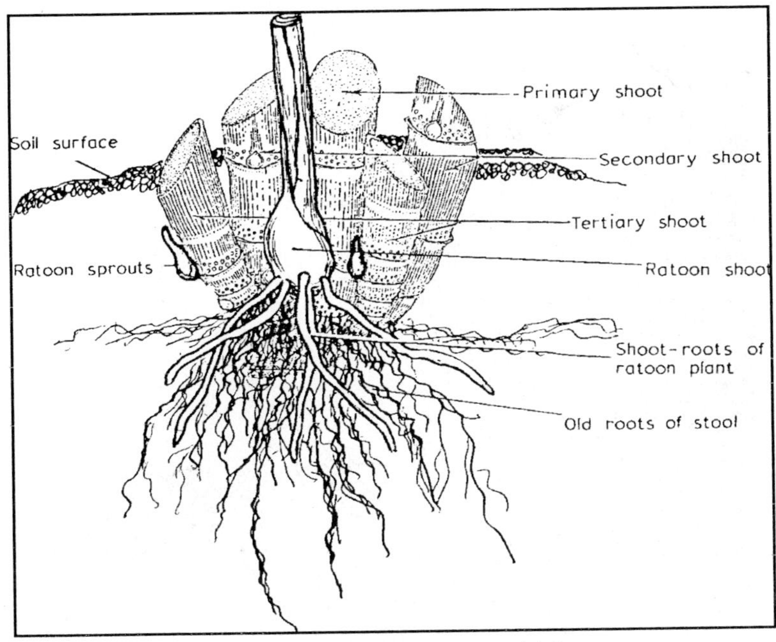

Fig. 2.1 : Sugarcane stubble

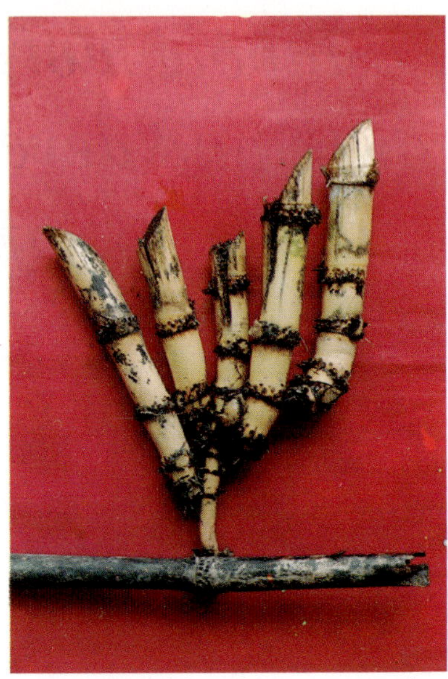

Fig. 2.2 : Sugarcane stubble after removal of root mass

Fig. 2.3 : Initial growth of plant crop (left) and ratoon (right)

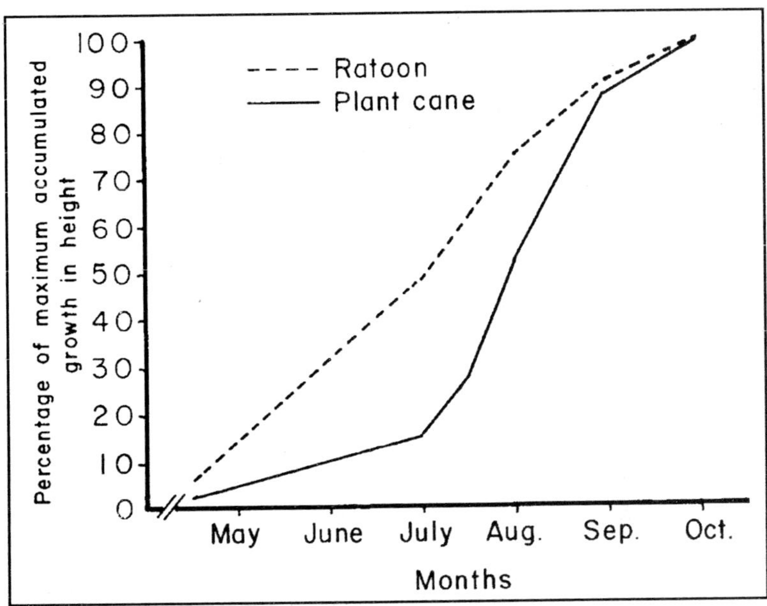

Fig. 2.4 : Percentage of maximum accumulated growth in height of plant and ratoon crops of sugarcane in different months

also with old root system of the previous crop. Stubble roots are dark brown in colour, hard and suberised. Sethiya and Dandsey (1992), at Karnal, studied the morphological parameters of a stubble, responsible for good ratooning, in nine sugarcane cultivars. They found that stubbles of good ratooners were profusely branched having smaller internodes with significantly more number of viable buds (Table 2.1). More number of viable buds per stubble is a desirable character for ratooning because these buds sprout to form shoots in ratoon crop. Stubbles of good ratooning cultivars also had comparatively more root biomass with deeper coverage.

Table 2.1. Number of branches and viable buds per stubble, internodes per branch, average internode length and weight of stubble roots per stubble in nine sugarcane cultivars of varying ratooning ability

Cultivar	Branches/ stubble	Viable buds/stubble	Internodes/ branch	Internode length (cm)	wt.of roots/stubble (g)	Length of roots (cm)	Ratooning ability
CoH 12	11.67±0.72	28.00±1.41	5.67±0.27	0.92+-0.06	466.17±35.75	65.93±2.29	Good
CoH 14	7.33±0.54	24.67±2.88	4.33±0.27	1.23±0.12	414.00±16.38	52.50±2.62	Good
CoS 767	5.33±0.54	18.67±1.96	5.00±0.47	1.03±0.06	298.54±18.35	39.53±1.75	Good
Co 1148	5.67±0.98	12.33±2.33	4.33±0.72	1.11±0.08	269.97±16.21	38.77±2.52	Average
Co 7717	3.67+-0.27	9.67±1.44	4.33±0.72	1.23±0.16	325.26±17.79	41.37±4.21	Average
Co 1158	4.33±0.27	6.00±0.27	4.00±0.47	1.32±0.17	221.19±35.07	37.29±1.36	Average
CoH 3	3.33±0.27	6.33±0.27	3.33±0.54	1.10±0.06	339.82±51.69	40.43±3.08	Poor
CoH 2	2.00±0.47	4.67±0.72	4.00±0.00	1.01±0.04	325.39±25.66	33.93±1.71	Poor
Co 7314	1.33±0.27	3.67±0.72	3.33±0.27	1.50±0.16	318.68±29.66	34.00±2.90	Poor

Source : Sethiya and Dandsey, 1992

2.2. **Shoot growth**

Initial growth of ratoon differs from plant crop (Fig. 2.3) because plant crop takes comparatively more time for emergence and initial establishment of plants while in a ratoon, stubbles possess an already established root system left from the previous crop. As a result emergence of shoots and their growth start almost immediately if the weather is suitable for growth. Ratoon shoots in initial stages, therefore, grow faster than the plant crop but the plant crop overtakes it in growth during *monsoon* season (Vasudeva Rao *et al.*, 1946, Raheja and Misra, 1954). In studies of Raheja and Misra (1954) at New Delhi, about 50 per cent of the cumulative shoot height had taken place by the beginning of *monsoon* season, compared to about 15 per cent in the plant cane under identical conditions of weather (Fig. 2.4). In the plant crop the accumulated growth in height during *monsoon* (between mid July to mid September) was about 70 per cent as against 43 per cent in the ratoon crop. In the post-*monsoon* the growth differences between ratoon and plant crops was less appreciable. Raheja (1946) reported that the ability of ratoon to develop shoots quickly and grow rapidly early in the season is a distinct advantage over the plant crop particularly in years with short formative and longer grand growth period.

2.3. **Root growth**

Initially, ratoons start functioning on the old stubble roots (Dillewijin, 1952). However, stubble roots are reported to be less efficient in absorbing nutrients and water since they are highly suberised. Hunsigi (1989) reported that the old roots gradually cease to function and decay and their place is taken by soft, succulent and white coloured new roots emerging from the base of the sprouting buds (Fig. 2.5). The transitional period between the decay of stubble roots and

Fig. 2.5. Old stubble roots (Black) and newly formed shoot roots (white)

the initiation of new shoot roots depends on the soil and climatic conditions prevailing at the time of ratooning and the cultural practices adopted. The period may vary from 4 to 6 weeks (Hunsigi,1985). After the death of old roots, the crop depends totally upon its shoot root system. However, there is an imbalance in shoot:root ratio at the initial stage because of the presence of comparatively very large number of shoots per clump with greater early growth. The extensive but less efficient stubble roots may fail to provide adequate nutrients and water to growing shoots. Since shoots of successive ratoons originate at a higher level than those of previous crop (Fig. 2.6) the root system of a ratoon crop is shallower than that of plant crop. Due to this phenomenon ratoons have less anchorage and are more prone to lodging and soil moisture stress. Shallow root system, however, confers an advantage which lies in the fact that ratoon is know

Fig. 2.6. Shoots of successive ratoons originating at a higher level.

to withstand waterlogging better than plant crop because of less adverse effect of waterlogging on the respiration rate of the ratoon crop than of plant crop (Khanna and Raheja, 1948).

2.4. **Tillering**

After harvesting the plant crop, the apical dominance of individual harvested cane is lost and the buds left on the stubble sprout and emerge as new shoots. Tillering in ratoons starts early and is more profused (Fig. 2.7) than plant cane (Raheja, 1956). It is because the left over stubbles in the ratoon contain more of initially available buds in a limited space than are available in a newly planted crop. In addition, ratoon crop has the advantage of the previous crop roots to

Fig. 2.7 : Tillering pattern in plant crop (left) and ratoon (right) during initial stage

start with. Therefore, buds of these stubbles germinate quickly to form a very high shoot density instead of tillers emerging from the plant sett to form a clump initially at a slow rate and progressively at a faster rate towards the later part of the summer season. Generally there are late under-developed shoots present at the time of harvest of sugarcane crop which are also cut down with the other canes leaving terminal growing points of many of these young shoots undamaged. These shoots get re-established and grow to form shoots quickly (Misra, 1953). More profuse tillering and quick production of tillers is a speciality of the ratoons. Ratoon started under favourable temperature and moisture conditions results in the production of about 80 per cent of the peak number of shoots within about a fortnight (Raheja and Misra, 1954). However, the tiller number per stool increases as the age of the ratoon advances (Fig. 2.8). The maximum number of tillers is generally attained in July. When the stools attain the maximum number of tillers, the mortality in tillers starts.

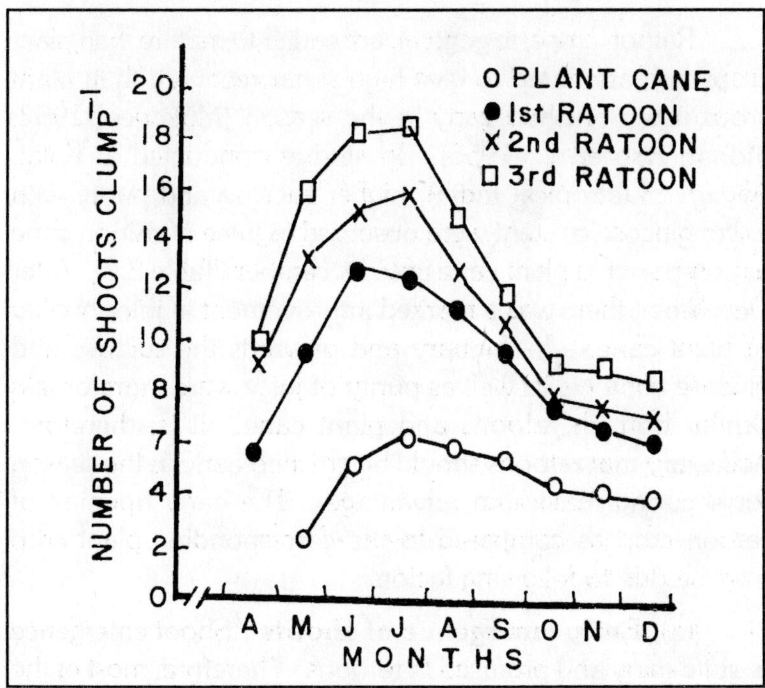

Fig. 2.8 : Production of shoots in plant and ratoon crops of sugarcane

The rate of mortality is very rapid till the month of October after which the number of canes per stool remains the same. Excessive tillering in the case of ratoon results in greater competition and consequently greater mortality than the plant cane. Number of shoots and millable canes increases gradually in subsequent ratoons. The average cane weight is relatively less in ratoon crops and it gradually decreases in subsequent ratoons. It is hypothesized that higher plant density of ratoon followed by high rate of mortality is probably responsible for reduced weight per cane.

2.5. **Ripening**

Ratoon crops, in general, are earlier to mature than plant crops and are noted to give high sugar recovery than plant crops when crushed early in the season (Mukerjee, 1952; Mohan Rao *et al.*, 1956). In studies conducted at Kalai, Aligarh (Subtropical India), higher sucrose and purity with lower glucose content were observed in juice of ratoon cane as compared to plant cane upto December (Table 2.2). After December, there was a marked improvement in juice quality of plant cane. In January and onwards the sucrose and glucose contents as well as purity of juice were more or less similar both in ratoons and plant cane. It is, therefore, necessary that ratoons should be crushed early in the season for securing maximum advantage. The early ripening of ratoon crop as compared to the corresponding plant crop may be due to following factors :

(a) **Early emergence of shoots :** Shoot emergence is quite early and profused in ratoons. Therefore, most of the millable canes at harvest come from these early formed shoots which are more in age than that of plant crop (Tandon and Misra, 1954). Panje (1965) also observed that early formed shoots were relatively superior to late formed shoots in quality characteristics.

(b) **Lower hydration :** Ratoon crop throughout the life cycle shows lower hydration of plant tissues as compared to the plant crops (Table 2.3). A close relationship was observed between earliness of cane and the moisture content

Table 2.2 : Juice quality and extraction per cent of plant and ratoon canes during maturity phase at Kalai (Aligarh) (Mean of data from 1942-43 to 1948-49)

Month	Plant cane	I Ratoon	II Ratoon	III Ratoon
Sucrose % in juice				
October	8.34	9.28	9.47	10.00
November	11.62	12.04	12.39	12.88
December	13.32	13.94	13.79	14.00
January	14.65	14.79	14.85	15.14
February	16.93	16.53	16.54	16.79
Glucose % in juice				
October	3.07	2.76	2.62	2.55
November	2.22	2.05	1.95	1.91
December	1.56	1.40	1.40	1.32
January	0.93	0.94	0.90	0.82
February	0.48	0.61	0.58	0.51
Purity % in juice				
October	65.6	70.2	71.1	72.1
November	75.7	77.8	78.4	79.3
December	80.4	82.0	81.7	81.9
January	84.4	85.2	84.9	85.4
February	88.3	87.2	87.3	87.8
Juice extraction % of cane				
October	66.6	65.2	63.7	61.6
November	66.1	64.3	63.9	63.4
December	66.2	65.0	63.6	63.0
January	66.1	64.8	64.1	63.7
February	61.9	60.2	58.9	58.3

Source : Anonymous, 1952

Table 2.3: Hydration levels of plant and ratoon crops of sugarcane (Co 419) at Anakapalle, Andhra Pradesh (Tropical India).

Month	Water content (%) in whole plant		
	Plant	I Ratoon	II Ratoon
July	80.7	79.9	-
August	81.2	79.6	80.2
September	80.6	78.5	75.1
November	74.9	73.8	73.5
January	72.6	69.1	73.1
February	72.1	67.9	68.2

Source :Modified from Mohan Rao *et al.* ,*1956*

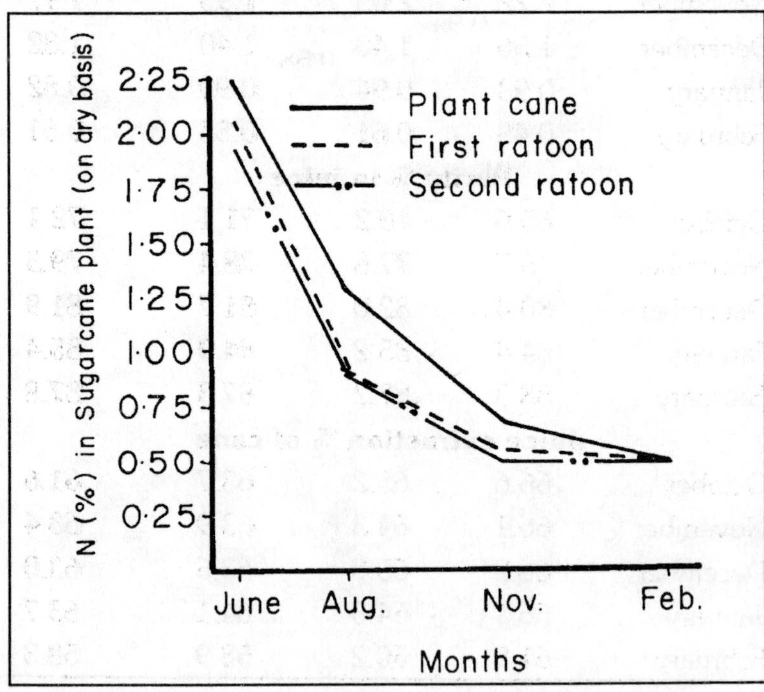

Fig. 2.9 : Nitrogen content in plant and ratoon crops of sugarcane

of tissues by Parthasarathy (1949). Moisture content in plant parts falls progressively as the cane matures. Thus, the lower hydration levels of the ratoons may be the possible cause for their early maturity.

(c) **Lower nitrogen content and its rapid fall :** The Nitrogen concentrations in the plant tissues remain lower in the ratoons than plant crop throughout the active growth period (Mohan Rao and Narasimhan, 1954). Besides, during grand growth, decline in N content in ratoons is steeper than the respective plant crop (Fig. 2.9). This characteristics appears to be responsible for superior juice quality of ratoons.

3. Agronomic practices in plant cane for good ratooning

3.1 **Method of planting :** Different methods of planting the plant cane appear to bear an important relationship with the yield of subsequent ratoon. In experiments conducted concurrently at different locations (Shahjahanpur, Muzaffarnagar and Gorakhpur) of Uttar Pradesh, trench planting (Fig. 3.1) yielded more in subsequent ratoon than that obtained from flat planting at Muzaffarnagar and Gorakhpur. However, at Shahjahanpur different methods of planting did not produce any significant difference (Tandon *et al.*,1957). The contradictory results may be attributed perhaps due to different fertility and soil moisture levels under

Fig. 3.1 : Trench method of planting sugarcane

different experiments. Singh and Kanwar (1965) working at Jalandhar (Punjab) reported that the trench planted plant cane produced better sprouting of subsequent ratoon. Michael Raj *et al.*(1982) also reported that trench planting of cane produced better ratoon as compared to flat bed or ridge and furrow systems in tropical India. Better ratoons from deep trench planted crops in comparison to shallow furrow planted crops are due to deeply located stubbles in the soil which results in better spread of root system.

3.2 **Time of planting :** Information related to the effect of time of planting of plant crop on subsequent ratoon performance is sporadically available. However, in studies conducted by Kumar and Srivastava (1993) at Karnal, yield of ratoons raised from plant crops of different planting times did not differ significantly (Table 3.1).

Table 3.1: Effect of time of planting of plant crop on cane yield of plant and ratoon crops of sugarcane at Karnal, Haryana (Subtropical India)

Planting date of plant crop	Cane yield (t/ha)		CCS% at harvest	
	Plant crop	Ratoon	Plant crop	Ratoon
15 March	81.2	58.8	10.5	10.4
15 April	76.6	64.4	10.7	10.1
15 May	65.0	61.5	10.4	10.5
CD (0.05)	4.3	NS	NS	NS

Source : Kumar and Srivastava, 1993

3.3 **Crop geometry and reed rate :** Optimum plant population arranged in a proper geometry is the important factor in obtaining higher yield of any crop. A considerable amount of work has been done on this aspect in relation to yield and quality of plant crop of sugarcane. And yet enough attention has not been paid to assess the impact of these different seed rates put in various row arrangements at plant cane stage on the performance of subsequent ratoon crop raised from them. Gill and Alam (1967) concluded that a reduction of 40 per cent in seed rate of autumn planted cane could be made without any adverse effect on the yield of subsequent ratoon at Lucknow. Likewise Shanmugasundaram *et al.* (1981) from Coimbatore reported that different seed rates of 90, 135, 180, 270 and 360 thousand buds per hectare did not make any significant difference in yield obtained from subsequent ratoons (Table 3.2). Mathur *et al.* (1968) also observed at Gorakhpur that high seed rate in double row planting (90:30:90 cm) did not improve the yield of subsequent ratoon crop as compared to conventional planting in rows spaced at 90 cm apart (Table 3.3).

Table 3.2 : Effect of different seed rates on cane yield of plant and subsequent ratoon crop of sugarcane at Coimbatore (Tropical India)

Seed rate ('000 buds/ha)	Cane yield (t/ha)	
	Plant crop	Ratoon
90	101.4	80.8
135	115.4	82.7
180	116.4	86.2
270	120.3	86.3
360	123.6	87.7
CD (0.05)	8.9	NS

Source :Shanmugasundaram, 1981

Table 3.3: Effect of planting patterns on cane yield of plant and subsequent ratoon crops of sugarcane at Gorakhpur, U.P. (Subtropical India) (Mean of data from 1962-63, 1963-64 and 1965-66)

Planting pattern	Cane yield (t/ha)	
	Plant cane	Ratoon
90:90:90 (Single row)	67.7	56.7
90:30:90 (Paired row)	66.9	51.7

Source :Mathur *et al.*, 1968

Note : One 3 budded sett per running 30 cm row length was planted in each case

4. Varieties for good ratooning

Selection of a variety is an important factor for good ratooning. The fact remains that a variety which gives higher yield of plant cane may or may not yield better in ratoons as well. Varieties with minimum rate of yield decline in successive ratooning as against plant crop are considered to be good ratooners. Co 650 a variety examplify it. During the plant crop performance trial at nine zonal centres in 1954-55 the variety Co 650 gave an average yield of 64.6 t/ha, but when ratooned at six centres it yielded to 37.0 t/ha only as compared to 55.5 and 50.5 t/ha by CoS 345 and CoS 321, respectively (Tandon *et al.*, 1957).

Ratooning ability of a variety is a genetic character and varieties can be classified as good/poor ratooners. Hybrid canes derive their capacity of better ratooning from *Saccharum spontaneum* in their parentage. Noble canes, on the other hand, are comparatively shy ratooner. However, the expression of ratooning is controlled by a strong variety-environment interaction. Appropriate external conditions alongwith favourable genes are necessary for exhibition of good ratooning. In case of early maturing varieties which as a rule are branded as poor ratooners, but not necessarily so. Ratooning in case of early maturing varieties in subtropical India is poor mainly due to low temperature at the time of ratooning.

In general, following points should be kept in mind while selecting a variety for better ratooning performance :

(a) Thicker varieties produce poor ratoons than thin and medium-thick varieties.

(b) Shy and slow germinating varieties are poor in ratooning than quick and better germinators. Good ratooners like Co 419, Co 1148, Co 1158, CoJ 64 and Co 7717 are good germinators too. Shanmugasundaram *et al.*, (1981) at Coimbatore (Table 4.1) found that a shy germinating variety Co 658 was poor in ratoon performance as compared to Co 419 and Co 6304 (good germinators). Similarly in another study conducted at Cuddalore (Tamil Nadu) from 1963 to 1970, variety Co 658 gave poorer ratoon than Co 419 and Co 449 because of its shy and slow germinating characteristics (Kadirvel, 1972).

Table 4.1: Yield performance of sugarcane varieties in plant and ratoon crops at Coimbatore, Tamil Nadu

Variety	Cane yield (t/ha)	
	Plant crop	Ratoon
Co 658	102.6	67.3
Co 419	119.8	86.1
Co 6304	125.6	109.9
CD (0.05)	NS	15.4

Source :Shanmugasundaram *et al.*,1981

(c) Varieties with minimum stubble deterioration are found to have better ratooning capacity.

(d) Deep rooted drought tolerant varieties have better ratooning capacity. The best ratooner varieties in subtropical region. *viz.*, Co 312 and Co 1148 are drought tolerant ones.

On the basis of large number of experiments and direct observations, varieties like Co 419 and 740 proved good ratooners in tropical India. In subtropical parts, varieties Co 312, CoS 510, Co 1148, Co 1158, CoJ 64, BO 17, BO 54 etc., were proved as good ratooners. BO 91, CoS 770, CoS 767 and CoS 802 are recent good ratooning varieties.

5. Cultural Practices

5.1 Proper time of starting ratoon

In general, the proper time of starting ratoons is the same which is considered as proper time for planting of plant crop in a particular region (Mathur, 1980). In subtropical India, February – March was found to be the optimum period (Fig. 5.1) for starting ratoon (Tandon *et al.*, 1955; Agrawal and Gupta, 1960; Bhoj and Mathur, 1965, Dixit and Bhardwaj, 1974). Ratoons from crops harvested either in the extremely cold winter months (December and January) or in extreme heat in April-May result in poor yield. There is a permanent adverse effect of low temperature on tillering capacity of ratoon if it is started during winter months. This

Fig. 5.1 : A field view of timely started ratoon crop

may be due to the damage of old roots owing to inactive life of stump during winter season and loss of nutrition through respiration and inversion. It may also be due to the damage of buds themselves before they find favourable temperature conditions for their development. If ratoon is started late in the season, say April-May, tillering is profuse but mortality is very high and growth of shoots is poorer in comparison to earlier started ratoon. It could be due to the shortage of time at the disposal of the crop to establish itself well to gain full advantage of the favourable weather conditions during rainy season for growth (Tandon and Misra,1954).

Because of more equitable climate (neither extremes of winter nor summer, as in subtropical India) in tropical India, differences in the time of harvesting are not likely to have a very significant effect on the yield of succeeding ratoon crop of the same duration, However, in late started ratoon crop, relatively shorter growth span of time reduces average cane weight. October to February is the optimum period of ratooning in tropical India, (Hunsigi, 1982).

5.2. **Post harvest field operations**

5.2.1 **Trash handling :** After harvesting sugarcane, handling of trash of the crop is very important in ratoon cropping. Leaf and stalk trash often litters the field after harvesting (Fig. 5.2). It causes difficulty in irrigation, fertilization, pest control and weed control as well as smothering young ratoon shoots buried beneath the trash layer. In India, sugarcane trash is either mulched or burnt in *situ* or removed from the field for thatch making, using as fuel and /or litter under animals in cattlesheds.

There are different schools of thoughts in trash management. In some cases trash is burnt (Fig. 5.3) to hurry up the ratoon operations. Trash burning ensures partial sterilization of soil and breaking the chain of carry over of insect pests and diseases (Tandon *et al.*,1957). Burning also kills the upper buds of the stubble and allows to sprout the

Fig. 5.2 : Presence of sugaracne dry leaves and stalks (trash) in the field after harvesting

deeper ones.

Many people do not recommend burning as it results in loss of a huge quantity of organic matter and suggest trash mulching (Fig. 5.4). Burning is only advisable when trash mulching leads to a build up of insect pests or diseases. Mulching has been known to affect sugarcane growth, and yield favourably because it checks water loss, suppresses weed growth, keeps soil in proper tilth even without hoeing and regulates soil temperature. Trash is also a potential source of organic matter in sugarcane cultivation. The trash spread acts as mulch in the current crop and after decomposition it

Fig. 5.3 : Trash burning in situ

increases soil organic matter which benefits subsequent ratoon crops by improving soil structure and nutrient recycling. The amount of organic matter and major nutrients present in

Fig. 5.4 : Trash mulching at ratoon initiation stage.

sugarcane trash are mentioned in Table 5.1.

Table 5.1 : Amount (Kg) of organic matter and major nutrients in per tonne of sugarcane trash and FYM.

Constituent	Trash	FYM
Organic matter	390.00	900.00
Nitrogen	5.30	5.40
Phosphorus	1.10	3.50
Potash	5.80	15.20

Further, trash mulching is convenient to adopt in ratoon cane as trash from the preceding crop could be made available in the field itself. To adopt mulching, trash should be collected and reapplied properly in between cane rows to form a 10 cm thick cover. Trash collected from one hectare sugarcane field (if it is 5-7 tonnes), is sufficient to give about, 10 cm thick cover over half hectare field. Results reported from experiments conducted at different locations under All-India Coordinated Research Project on Sugarcane showed definite superiority of trash mulching in improving the yield of ratoon crop over other methods of trash disposal (Anonymous, 1974).

5.2.2 **Dismantling of ridges** : Deep furrow planting followed by heavy earthing up buries 25-30 cm of millable cane length in the soil, and at least 15-20 cm of buried cane can be harvested by dismantling the ridges. A gang of 15-20 labourers may be employed to break open the ridges prior to harvesting. This will enable harvesting canes at least 15-20 cm below its usual point. The addition of 15-20 cm length of cane will amount to 5-6 tonnes of additional yield of cane per hectare, and at the same time, no operation like stubble shaving will be necessary. Other operations for preparing land can be done after removing the trash as early as possible.

5.2.3 **Stubble shaving :** The level from which the shoots emerge at the start of the ratoon crop is very important. The anchorage provided by the deeper roots is more effective and firm than when the root growth is at a higher level near the surface of the soil. In addition, deeper and well spread roots account for better foraging capacity and better utilization of soil moisture and nutrients resulting in better crop growth (Misra and Mathur, 1983). The height above the ground level at which stalks of plant crop (or preceding crop) is cut has an important bearing on the level from which shoots will emerge (Fig. 5.5). Due to top dominance upper most bud present on stubble sprouts first. In case it is located above the ground, shoot roots are unable to enter the soil and a small plant die. Sometimes, such high placed sprouts initially depend on the root system of old stubbles that are unable to grow fresh roots rapidly

Fig. 5.5 : Leaving upper ground portion of cane unreaped leads to shallow root system

and thereby suffer in vigour. Hence, care is required in harvesting the crop flush to the ground level. This will stimulate the growth of the lower buds with deeper root system. However, in practice very often crop is harvested leaving stumps ranging from 5 to 10 cm above the ground. It causes appreciable loss of cane yield and also depresses the growth and yield (Table 5.2 and 5.3) of succeeding ratoon (Chauhan, 1993). Stubble shaving should be done to remove superficial buds if harvesting has not been done on ground level. Vaidya (1965), Agrawal (1981) and Paneerselvan *et al.*(1992) observed the increased growth and yield of ratoon crop due to stubble shaving as compared to no stubble shaving. However, the response to stubble shaving can only be evident with careless harvesting of the previous crop leaving long butts above the ground (Singh *et al.*, 1975).

Stubble shaving should be done before they sprout. Late stubble shaving allows the stubbles to sprout and some energy and food material is wasted, and then ratoon depends on the second sprouting of the deep placed buds. The second sprouting may not be so even, healthy and strong as the first ones. Sometimes, stubble fails to give out new sprouts if shaved late. This leads to weak and less number of millable canes in the ratoon crop (Adhate, 1960).

Stubble shaving should be done with a sharp edged spade (*phaora*) or *pankati*. A spade which is blunt, shakes the stubble and disturbs the root system which is not conducive to proper ratooning. Blunt spade may also cause bruises to the stubble. The bruises to stubbles result in the fermentation of the sucrose in the stubbles and then the stubbles die away. For efficient and timely operation at much less cost, a tractor mounted rotary stubble shaver (Fig.5.6) designed by the Indian Institute of Sugarcane Research, Lucknow can be used.

Table 5.2: Effect of height of plant cane harvesting and stubble shaving on number of millable canes and yield of ratoon crop at Lucknow (Subtropical India)

Treatment	Millable canes (000/ha)			Cane yield (t/ha)		
	1986-87	1987-88	Mean	1986-87	1987-88	Mean
A. Harvesting of plant crop						
(i) At ground level	85.7	94.9	90.3	58.0	70.0	64.4
(ii) At 5 cm above ground level	80.0	85.7	82.9	50.9	64.4	57.7
(iii) At 10 cm above ground level	78.5	83.9	81.2	47.5	62.0	54.8
CD (0.05)	5.3	7.1	-	5.9	5.1	-
B. Stubble shaving						
(i) No stubble shaving (usual harvesting)	85.6	95.0	90.3	64.1	74.9	69.9
(ii) Stubble shaving	78.8	85.0	81.9	54.0	64.0	59.0
CD (0.05)	5.3	7.3	-	6.3	6.2	-

Source: Chauhan, 1993

A bullock drawn stubble shaver has also been developed.

Table 5.3. Effect of stubble shaving on shoot population and yield of ratoon crop at Melalathur, Tamil Nadu Tropical India)

Treatment	Cane yield (t/ha)	Shoot (000/ha)			Millable canes (000/ha)
		20	50	90	
		days after stubble shaving			
No stubble shaving	88.8	131	217	174	74.3
Stubble shaving	101.6	173	252	188	86.6
CD (0.05)	1.4	7.5	4.6	5.4	1.2

Source: Modified from Paneerselvan *et al.*,1992

Fig. 5.6 : A tractor mounted rotary stubble shaver

5.2.4 **Gap filling :** Presence of properly spaced three viable clumps per running meter row length is considered to be a normal stand for ratoon crop (Prasad *et al.*, 1981). Reduction in three viable clumps per running meter causes considerably long vacant space within the row. These vacant intra-row spaces may be called as gaps. The gaps in ratoon fields reduce the yield by lowering the millable cane population. Gaps in ratoon are the result of either poor stand of plant crop due to reduced germination or/and loss in capacity of stubble buds of previous crop to sprout. Stubble buds may lose their capacity to sprout because of low temperature at ratoon initiation, physical injury to stubbles at the time of harvesting of previous crop and attack of insect pests and diseases. In subtropical India, buds located on the stubble of winter (December-January) harvested crops remain dormant during this low temperature period and sprout only when favourable temperatures in February-March are available. By this time, the capacity of the stubble buds of low temperature harvested crop is appreciably reduced. Therefore, time of harvesting of previous crop profoundly affects the intensity of gaps in following ratoon. Studies at Karnal indicated that harvesting the plant crop in winter months (December –January) caused 18 to 26 per cent gaps as against 11 to 13 per cent in spring (February) harvested crop (Mathur, 1986). At Lucknow, harvesting plant crop in winter (December) resulted in 22 to 30 per cent gaps in succeeding ratoon crop as compared to 10-17 per cent in spring (February) and late spring (April) harvesting (Chauhan and Motiwale, 1989).

Varieties also differ significantly in failure of clumps and occurrence of resultant gaps. Misra and Mathur (1990) at Karnal, observed 19.6, 27.9 and 18.2 per cent gaps respectively in three early varieties of canes *viz.* CoJ 64, Co 7314 and Co 7717 harvested in first week of January.

Number of gaps also increases progressively with the increase in the number of ratoons from a single planting. Mathur (1986) recorded 23.9, 34.9 and 48.1 per cent gaps respectively in first, second and third successive ratoons started in December at Karnal, where gaps remained unfilled.

Gappy stand give poor ratoon yield. However, reduction in 10 per cent clump population does not cause significant loss in cane yield (Prasad *et al.*, 1981) due to compensatory ability of ratoon crop for space (Raheja and Misra, 1954). But when the gaps are more than 10 per cent, corresponding reduction in number of millable cane stalks and yield of ratoon cane are observed. At Lucknow, decrease in cane yields recorded were to the tune of 1.5, 9.9, 15.9, 23.0, and 33.0 per cent when intensity of gaps created were 10, 20, 30, 40, and 50 per cent, respectively, in first ratoon of sugarcane variety Co 1148.

If gaps exceed more than 10 per cent of the total clump population, gap filling is done to attain adequate plant population. Gaps are filled either by transplanting ratoon stubbles of the same variety or by settlings raised either in nursery beds or in polythene bags (Fig. 5.7) (Tandon *et al.*, 1957; Panje, 1967; Prasada Rao, 1980; Panje 1964; Bhoj, 1976 and Rao, 1981). Sometimes fresh sugarcane setts are also used. Fresh setts planted in gappy areas take quite a long time to germinate and plants coming up from the setts remain comparatively shorter than the ratoon plants which have already developed much by the time setts germinate. The bigger ratoon plants do not allow the smaller ones to develop much and thus loss in yield is not made up substantially.

Ratoon stubbles of the same variety may be taken from other field which is proposed to be ploughed up. Raising settlings in nursery consists of planting setts in nursery beds,

Fig. 5.7 : Polythene raised settling (left), nursery bed raised settling (middle) and ratoon stubble slip (right)

and allowing them to sprout and grow for some time. Nursery is better located near to the source of irrigation for convenience. Settlings are most suited for transplanting in about 7 to 9 weeks after planting. If the leaves of the settlings are found to be long, they can be cut off at half length before transplanting to minimise transpiration and consequent wilting before they get established. Kanwar and Sharma (1981) advocated raising of settlings in polythene bags for quick and better establishment as compared to settlings raised in nursery beds. Nursery raised settlings are prone to high mortality because the root system of the young shoots is disturbed during their uprooting from nursery beds. A single bud cutting of sugarcane stalk is placed in polythene bag of one kilogram capacity filled with soil for raising settlings. After germination the settling is transplanted with earth ball by removing polythene.

Gap filling should be done in early tillering phase (February-March) as delay in gap filling does not keep pace with normal shoots in growth and development, and are many time smothered out. Soil of the area for gap filling should be worked out well before placing settlings/stubbles. Thorough irrigation is given soon after gap filling to facilitate better establishment of transplanted material.

Multi-location field experiments conducted through All India Coordinated Research Project on Sugarcane also indicated the need for gap filling in ratoons for increasing sugarcane productivity. However, the range in yield increase (7.3 to 40.1 per cent) due to gap filling was more in subtropical India (Table 5.4) as compared to (2.2 to14.4 per cent) tropical belt of India (Table 5.5). It is because of the fact that in tropical India, crop does not suffer from extremes of cold and heat. Moreover, in this belt germination per cent of cane buds planted is, in general, considerably high than that obtained in sub-tropical India. Thus, the effective clump population per hectare in this region is much higher than in subtropical belt.

Table 5.4: Effect of gap filling on ratoon yield (t/ha) in subtropical India

Treatment	Jalandhar		Karnal		Pantnagar		Pusa
	84-85	85-86	84-85	85-86	83-84	85-86	84-85
No gap filling	38.9	43.2	53.5	55.9	46.6	46.4	49.9
Gap filling	54.5	56.5	59.1	60.0	48.9	54.6	59.2
CD (0.05)	2.6	2.2	2.6	0.5	NS	2.5	6.3

Table 5.5 : Effect of gap filling on ratoon yield (t/ha) in tropical India

Treatment	Thiruvella		Navsari		Cuddalore		Sehore	Padegaon	
	83-84	84-85	83-84	84-85	84-85	85-86	84-85	85-86	85-86
No gap filling	118.3	81.3	71.7	60.9	95.4	98.4	76.8	65.7	80.5
Gap filling	122.9	85.0	83.8	66.3	98.8	102.1	78.5	67.7	85.7
CD (0.05)	NS	NS	1.9	5.0	1.7	1.9	0.6	1.2	NS

Source: Kanwar *et al.* (1988), Verma *et al.*(1987), Patel et al.(1989), Pandian and Rajsekaran (1991) and Tech. Rep. (Agronomy) AICRP on Sugarcane 1983-84, pp. 71-75; 1984-85, pp. 20-26 and 1985-86, pp. 26-30.

5.2.5 **Removal of compaction :** Compaction of the soil is a major disorder with ratooning. The loose and friable soil at the beginning of the plant crop gradually becomes firm and compact at the commencement of ratoon crop. Constant trespassing of men and machinery contribute to compaction particularly in the root zone. Compaction affects several soil physical characteristics such as bulk density, pore-space, structure and hydraulic and thermal properties of soil. Observations under alluvial loams of *Gangatic* belt have indicated that with successive ratooning in the same field, the soil in the root zone has a tendency to get compacted. The bulk density of soil at different depths significantly increases with successive ratooning from first to third ratoons (Table 5.6). The compaction interferes with normal and healthy plant growth. Roots of the ratoon plants experience inadequate soil aeration due to reduction in macro-pores. This tends to decrease intake of water by soil and restricts water and nutrient uptake by roots. Basically roots can penetrate compacted soils up to a critical bulk density which is specific for specific soil materials. Singh and Lal (1963) found negative correlation of –0.73 and –0.59 between bulk

Table 5.6: Bulk density (g/cm^3) and pore space (%) of alluvial loam soil at various depth as influenced by successive ratooning of sugarcane at Lucknow (Subtropical India)

Depth	Average bulk density		
(cm)	I ratoon	II ratoon	III ratoon
0-15	1.49(43.8)	1.59(40.0)	1.61(39.2)
15-30	1.57(40.8)	1.61(38.9)	1.63(38.5)
30-45	1.63(38.5)	1.65(37.7)	1.73(34.7)
45-60	1.67(37.0)	1.67(37.0)	1.76(33.5)

Figures in parentheses are % pore space

Source: Srivastava, 1981

density of soil and dry weight of roots and shoots of sugarcane respectively (Table 5.7). Srivastava (1983) also observed a negative response of sugarcane ratoon yield and growth parameter with increasing level of compaction (Table 5.8). A

Table 5.7: Effect of bulk density of alluvial loam soil on the weight of roots and shoots of sugarcane at Lucknow (Subtropical India)

Bulk density (g/cm^3)	Dry weight of roots (g)	Dry weight of shoots (g)
1.1	1.2	2.6
1.2	1.2	2.3
1.3	0.9	2.0
1.4	0.9	1.6
1.5	0.7	1.6
1.6	0.3	1.5

Source: Singh and Lal, 1963

Table 5.8 :Effect of soil compaction on the growth and yield of sugarcane ratoon raised on alluvial clay loam soil at Lucknow (Subtropical India)

Yield and yield component	Compaction level as bulk density (g/cm^3)						
	1.34	1.45	1.53	1.60	1.66	1.71	CD(0.05)
Cane height (m)	2.04	1.88	1.66	1.64	1.57	1.52	0.11
Cane girth (cm)	2.14	2.02	1.92	1.90	1.88	1.86	0.06
Cane yield (t/ha)	73.6	59.8	54.7	46.8	46.9	47.1	5.5

Source: Srivastava, 1983

highly significant negative correlation (r= –0.92) was obtained by him between stalk yield and bulk density of soil and critical level of compaction resulted in bulk density range of 1.51 to 1.54 g/cm^3 for clay loam soils of Subtropical India.

In order to have adequate development of the fresh roots, the soil must be loosened. This is done by irrigating the field immediately after the harvest of preceding crop and loosening the soil by suitable implement at an appropriate moisture content. Sub-soiling is reported to be one of the best means of increasing aeration in a compact or heavy soil. Prasada Rao *et al.* (1979) observed that deep ploughing in black soils increased ratoon yield by 7.5 per cent at Anakapalle (Andhra Pradesh). However, in red sandy loam soils of Mandya, Karnataka (Hunsigi, 1982) and alluvial sandy loam of Lucknow (Chauhan and Singh, 1986), ratoons did not respond to deep plouging. This is possibly due to light texture which does not compact as much as black soils.

Another practice is "off-barring" in which plough or disc is used to cut away the soil on either side of the clump and to till the compacted inter-row for aerating the soil. The old root system is severed by this operation allowing fresh roots to emerge quickly. It ensures better utilization of water and applied nutrients.

5.2.6 **Weed control :** Weeds offer strong competition for growth resources like nutrients, sunlight, water and space with sugarcane plants. Thus, control and management of weeds is as important in the ratoon crop as in plant crop. Yield losses from 25 to 50 per cent (Srivastava and Kumar, 1992) in ratoon cane have been observed due to weeds. Weed problem in ratoon is of different dimensions than plant cane i.e. (i) where ratooning is practiced with delayed or no irrigation in the initial stage, weed establishment is usually poor, (ii) where irrigation is given to promote better establishment of the ratoon crop, the subsequent inter-cultural operations like hoeing or off-barring of inter-row spaces also delay germination and build up of the population of weeds, (iii) trash burning is another practice which confers better capacity to the ratoon crop to suffer less from weed competition, and (iv) prior establishment of stubble and their roots promotes fast initiation of tillering unlike slow and delayed germination in case of plant crop. Therefore, in ratoons the "rapid-close-in" is much earlier than plant crop. Hence the desirable weed-free environment in ratoon fields is the first 10-12 weeks after its initiation.

Most commonly occurring weeds in ratoon fields are *Cyperus rotundus L., Cynodon dactylon Pers., Solanum xanthocarpus Syn. Solanum nigrum L., Acalypha indica L., Convolvulus arvensis L., Convolvulus microphylla L., Dactyloctenium aegyptium (L), Digitaria sanguialis (L), Digera alternifolia (L.), Digera arvensis Forsk., Eclipta prostrata(L.), Eclipta alba Hassk, Euphorbia hirta L., Euphorbia thymifolia L., Portulaca oleracea L., Launes fallax Kwntz., Trianthema portulacastrum L., Trianthema monogyna L., Phyllanthus niruri L., Amaranthus spinosus L., Amaranthus viridis L., Cleome viscosa L., Corchorus trilocularis L., Corchorus acutangulus Lamk., Ipomea hispida Roem & Sch., Ipomea pestigridis* L. etc.

Various methods, generally, employed for controlling weeds in sugarcane ratoon fields may be broadly grouped as (a) cultural, (b) chemical, (c) mulching and (d) integrated approach.

Under cultural methods, tillage operations are done either manually or by bullocks or tractor drawn implements which not only eliminate weeds but also stir up the soil to increase soil aeration. For manual hoeings, different hand operated implements such as *kassi*, *kasola*, *spade*, hoses etc. are used. Various types of cultivators, hoes and harrows, including blade harrow (*bakhar*) are the common bullock-drawn implements in use for weeding in sugarcane. Tractor-drawn implements used include cultivators, rotavators and harrows. Recently a tractor mounted PTO shaft-operated multipurpose implement has been designed and developed at the Indian Institute of Sugarcane Research, Lucknow. With the help of this implement, all the initial cultural operations namely tillage in the inter-row spaces, stubble shaving, fertilizer/or insecticide application may be carried out simultaneously in one pass.

Trash mulching is an efficient and economical method of controlling weeds in sugarcane fields. In ratoon crops, weed control through trash mulching may be more convenient as the trash is available from the previous crop in the field itself. Experiments conducted at various sugarcane research stations have shown that spreading a 10 cm thick layer of sugarcane trash in ratoon fields immediately after initial operations (stubble shaving, off-barring and fertilizer application) suppresses most of the weeds effectively. It also eliminates subsequent hoeings because trash cover keeps the soil in proper tilth even without hoeings. In addition, trash mulch conserves soil moisture which increases its utility under limited availability of irrigation water (Table 5.9). Trash collected from a two hectare sugarcane field (10-14 tonnes), is sufficient to give about 10 cm thick cover over one hectare field. Although the benefit of trash mulching in ratoon fields has been reported from almost all the experiments conducted at different locations, care has to be taken to see that trash does not carry any disease or hidden insect pests. In case, there is subsequent possibility of white ant infestation, application of BHC 5 per cent dust at the rate of 20 kg per hectare may be adopted with advantage.

Table 5.9 : Comparison of manual hoeings with trash mulching for controlling weeds in sugarcane ratoon crop under deficient irrigations at Gorakhpur, Uttar Pradesh (Subtropical India)

Treatment	Cane yield (t/ha)			
	1963-64	1964-65	1965-66	Mean
Manual hoeings	54.6	57.4	42.4	51.5
Trash mulching	59.4	73.5	51.9	61.6
CD (0.05)	-	10.4	6.4	-

Note: (a) 10 cm thick layer of trash mulch was applied during April

(b) Only two pre-*monsoon* irrigations were given to crop

Source: Mathur and Singh, 1968

Triazines particularly atrazine and metribuzin (2 kg a. i./ha) applied at ratoon initiation as pre-emergence treatment (in relation to weeds) have been found quite effective in controlling weeds in ratoon fields. Most of the weeds occurring in sugarcane fields are effectively controlled by these herbicides. However, *Cyperus rotundus* is least affected with these treatments. A post-emergence application of 2,4-D (0.5 kg a. i./ha) could yield the desirable results. Since physical manipulation of soil at the ratoon initiation is essential which also controls weeds, an integrated method of weed control combining herbicidal applications with initial tillage operations proved more effective in controlling weeds and increasing cane yield (Table 5.10).

Table 5.10 :Weed control efficiency (WCE) of different cultural-cum-chemical methods of weed control and their effect on sugarcane ratoon yield

	Treatment	Cane yield (t/ha)	WCE
T1.	Atrazine @ 2.0 kg a.i./ha at the start of ratoon+2,4-D @ 0.50 Kg a.i/ha as post-emergence treatment	60.7	76.3
T2.	Metribuzin @ 2.0 kg a.i./ha at the start of ratoon+2,4-D @ 0.50 Kg a.i/ha as post-emergence treatment	60.7	77.6
T3.	One intercultivation followed by T1	65.4	79.8
T4.	One intercultivation followed by T2	65.1	79.1
T5.	Conventional method of manual hoeings	66.6	93.5
T6.	Unweeded check	42.2	

Source: Srivastava and Kumar,1992

5.2.7 **Earthing up** : Some varieties like CoS 767, Co 419 and Co 6806 tiller heavily with high rate of mortality. Heavy early (at two to two and half months of crop stage) earthing up ensures adequate plant population with less tiller mortality.

This operation promotes the increase in girth, height and weight of the millable canes. The quality of cane juice from such ratoons is also superior since there are few surviving late tillers. Heavy early earthing up also favours the formation of adventitious roots or the 'buttress' roots which provide good anchorage (Hunsigi, 1982).

6. Improving stubble sprouting in winter started ratoons

In Subtropical India, ratoon started from winter (December-January harvested sugarcane crop yields low (Tandon *et al*., 1957) due to lower plant population caused by poor sprouting of subterranean buds. These subterranean buds remain dormant during low temperature period and start sprouting only with the rise in temperature during spring (February-March). By this time the potentiality of bud to germinate is greatly reduced. It is a major problem for increasing ratoon productivity. It also adversely affects the adoption of autumn planting and early maturing high sugared varieties in Subtropical India as it necessitates harvesting in early crushing period which coincides with peak of winter season. Multiple ratooning is also restricted because the harvesting of preceding ratoon crop falls in colder months and therefore succeeding ratoon faces adverse effect of low temperature during its starting phase. Efforts have been made to overcome the effect of low temperature by using some plant hormones, polythene mulch or intercropping of certain *rabi* crops.

6.1 **Use of plant hormones :** Spraying of plant hormones like ethrel, IBA, TIBA, GA, Cycocel, IAA, NAA, ethephon or kinetin on stubbles were found to improve the sprouting of subterranean buds significantly which consequently enchances the yield of winter started ratoons. However, application of hormone (ethrel) with pesticide (gamma HCH or Emisan) produced significantly better effects than spraying of hormones alone. Ethrel + Urea, GA, IAA and NAA were also effective in enhancing stubble bud sprouting when sprayed on preceding standing sugarcane crop 15 days before its harvesting. Efficacies of different hormonal treatments in increasing yield of ratoons started during colder months are given in Table 6.1. Ethrel proved the best among the different plant hormones. Treatment with ethrel causes the release of ethylene in plant tissues (Warner and Leopold, 1969) which stimulates peroxidase activity leading to the destruction of endogenous auxins responsible for the inhibition of buds. In studies of Singh and Kanwar (1991) level of soluble proteins and sugars as well as activity of nitrate reductase and invertase in the subterranean buds were significantly enhanced due to ethrel application (Table 6.2).

Table 6.1: Efficacy of different hormonal treatments in increasing yield of ratoons started during colder months in Subtropical India

Effective treatment	yield (t/ha) increase	Reference	Location
(A)Sprayed on stubbles just after harvesting of preceding crop			
Ethrel (500ppm)	21.0	Kanwar and Kaur (1977)	Jalandhar
IBA (100ppm)	18.0		
ITBA (50ppm)	15.2		
GA (100ppm)	14.3		
Ethrel (500ppm)	10.4	Kanwar and Kaur (1981)	Jalandhar
GA (100ppm)	10.3		
Cycocel (5 l/ha)	9.9	Chauhan *et al.*(1984)	Lucknow
Ethephon Ethrel	14.1	Singh and Kanwar (1988a)	Jalandhar

(250ppm)			
Ethephon	14.1	Singh and Kanwar (1989)	Jalandhar
Cycocel (5 l/ha)	7.0	Verma and Yadav (1988)	Lucknow
Ethrel (500ppm) + Emisan(0.25%)	27.5	Singh *et al.* (1988)	Jalandhar
IAA (100ppm)	3.4	Singh (1990)	Jalandhar
(B) Sprayed 15 days before harvesting on preceding crop			
GA (50ppm)	10.9	Sharma *et al.* (1985)	Jalandhar
IAA (150ppm)	9.4		
NAA (200ppm)	6.5		
Ethrel (200ppm) + Urea (4% solution)		Sharma *et al.* (1987)	Jalandhar

Table 6.2: Protein level (mg/g fresh weight), nitrate reductase activity (mg nitrate liberated/g fresh weight/hr), soluble sugar level (mg/g fresh weight) and invertase activity (mole reducing sugar/ml enzyme extract) in subterranean buds under different hormonal treatments

Hormonal treatment	Soluble protein	Nitrogen reductase activity	Soluble sugar	Invertase activity
Ethrel (500 ppm)	10.74	0.38	18.2	1.80
IBA (150 ppm)	9.68	0.35	17.4	1.76
Kinetin (20 ppm)	8.24	0.34	14.8	1.59
GA3 (110 ppm)	8.04	0.32	14.5	1.62
Control	6.56	0.22	11.4	0.96
CD (0.05)	3.27	0.12	5.31	0.75

Note : Stubbles were sprayed with plant hormones immediately after harvesting of plant crop on 25th November

Subterranean buds were analysed after 15 days of treatment

Source : Modified from Singh and Kanwar, 1991

6.2 **Mulching** : Covering stubbles, during winter, with transparent polythene sheets with their ends tucked in soil increased sprouting of stubble buds apparently due to high temperature inside the polythene cover whereas spreading of black polythene and organic mulches (sugarcane trash, paddy straw, paddy husk and farmyard manure) exerted very severe adverse effect on sprouting and yield (Kanwar and Kaur, 1977 and 1981; Chauhan *et al.*, 1984).

6.3 **Intercropping** : Winter started ratoon when intercropped with potato produced higher number of millable canes with greater thickness and length (Table 6.3). Profuse stubble bud sprouting which resulted into higher tiller production throughout the tillering phase in ratoon with potato

Table 6.3: Crop yields, length and girth of cane with juice quality as influenced by different ratooning systems at Lucknow (Subtropical India)

Ratooning system	Yield (t/ha)		Cane length	Cane girth	Juice quality	
	Cane	Intercrop	(cm)	(cm)	pol%	purity%
December started ratoon (Sole)	48.0	-	175	2.20	18.3	91
December started ratoon+potato	62.0	13.3	201	2.28	18.3	92
December started ratoon+wheat	43.5	3.6	155	2.10	18.4	92
February started ratoon (Sole)	59.0	-	195	2.26	18.8	92
CD (0.05)	4.7	-	12	0.10	NS	NS

Sugarcane (Cv Co 1148), potato (Cv Kufri chandramukhi) wheat (Cv K 816); row spacing for sugarcane 90 cm; row ratios (1:1 for Sugarcane:Potato and 1:3 for Sugarcane:Wheat)

Source: Verma and Yadav, 1988

Table 6.4 : Tillers and millable canes (000/ha) under different ratooning systems at Lucknow (Subtropical India)

Ratooning system	March	April	June	July	Tillers Millable
December started ratoon (Sole)	72	121	184	129	90
December started ratoon + potato	137	219	236	144	108
December started ratoon +wheat	33	35	170	111	86
February started ratoon (Sole)	129	220	237	145	106
CD (0.05)	19	27	23	8	6.1

Source :Verma and Yadav, 1988

Table 6.5: Economic evaluation of different ratooning systems

Ratooning system	Cost of cultivation (Rs./ha)			Net profit in (Rs./ha)		
	Ratoon	Intercrop	Total	Ratoon	Intercrop	Total
December started ratoon (Sole)	4600	-	4600	5480	-	5400
December started ratoon + potato	4600	5250	10120	8420	2460	10880
December started Ratoon + wheat	4600	2720	7320	4535	4390	8925
February started ratoon (Sole)	4600	-	4600	7790	-	7790

Selling price per tonne of potato Rs. 600, sugarcane Rs. 200, wheat grain Rs. 1,600 and wheat *Bhusha* Rs. 250

Source: Verma and Yadav,1988

(Table 6.4) showed the beneficial effect of potato for bud sprouting in winter ratoons. Frequent irrigations alongwith liberal nitrogen application in potato might have promoted the physiological activities of sugarcane buds by increasing hydration status and soluble nitrogen contents of buds which in turn resulted into profuse sprouting. Potato intercropping with winter ratoons was also found highly profitable (Table 6.5).

7. Nutrient Management

Ratoon follows the preceding cane crop on the same soil, which remained for about a year or so. As sugarcane is a heavy feeder of nutrients, the soil available for ratoon crop gets depleted of the same. A 75 tonne cane crop absorbs 96 kg N, 54 kg P_2O_5 and 199 kg K_2O per hectare from soil (Adhate, 1960). However, the quantity may vary in different agroclimatic regions with different varieties but it is widely accepted that the quantity of phosphorus is more or less half of nitrogen, while potassium quantity is about double that of nitrogen. It is, therefore, necessary that in order to have good yields from ratoons, they are given adequate quantities of needed nutrients through manures and fertilizers. The doses of fertilizers and manures will depend upon soil type, soil organic matter and nutrient contents, growth period of crop and other factors.

7.1 **Response to nitrogen :** Response of sugarcane to applied N is explicit and nation wide. It has been found that ratoon crop generally needs 25 to 50 per cent higher level of nitrogen than plant cane (Seetharamaiah *et al.*, 1952; Mohan Rao and Narasimhan, 1954; Tandon *et al.*, 1955; Singh *et al.*, 1977; Subramanian, 1982; Tangaraju, 1982; Prasad *et al.*, 1984 and Mathur, 1987). Based upon the fertilizer trials, the N rate broadly varies from 120-225 kg per hectare for subtropical India and 224 to 400 kg per hectare for tropical India (Table 7.1). Yield responses of ratoon to per kg applied N/ha at recommended doses were recorded as 254 kg at Anakapalle (Lakshmikantham, 1973); 247 kg at Kalai, Aligarh (Mukerjee, 1952); 215 kg at Kanpur (Rathi and Singh, 1982), 165 kg at Shahjahanpur; 160 kg at Muzaffarnagar (Tandon

Table 7.1 : Nitrogen doses (Kg/ha) up to which significant increase in ratoon cane yield was recorded at different locations of India

Location	Significant yield increase up to (KgN/ha)	Soil type	Reference
Subtropical India			
Shahjahanpur (U.P.)	120	Alluvial (Sandy loam)	Tandon & Misra (1956)
Shahjahanpur (U.P.)	170	-do-	Bhoj (1962)
Lucknow (U.P)	140	-do-	Kapoor (1952)
Lucknow (U.P.)	200	-do-	Singh & Sharma (1973)
Kanpur (U.P.)	160	-do-	Jafri *et al*. (1978)
Pantnagar (U.P)	150	Clay loam	Singh *et al*.(1981)
Jalandhar (Punjab)	140	Alluvial (Sandy loam)	Rattan & Singh (1954)
Jalandhar (Punjab)	225	-do-	Kanwar & Kaur (1981)
Jagadhari (Haryana)	225	-do-	Singh *et al*. (1977)
Sriganganagar (Rajasthan)	200	Sandy Loam	Siag *et al*.(1989)
Pusa (Bihar)	150	Calcareous alluvium	Prasad *et al*.(1981)
Pusa (Bihar)	150	-do-	Thakur *et al*.(1986)
Tropical India			
Anakapalle (A.P.)	224	Sandy loam to clay loam	Lakshmikantham (1973)
Anakapalle (A.P.)	224	-do-	Prasada Rao *et al*. (1981)

Godavari (A.P.)	300	-	Vaidya (1965)
Navsari (Gujarat)	250	Clayey soil	Joshi *et al.*(1981)
Cuddalore (Tamil Nadu)	337	Coastal alluvium (Sandy loam)	Kadirvel (1972)
Coddalore (Tamil Nadu)	375	-do-	Daniel & Sanjivi Rao(1980)
Nellikuppam (Tamil Nadu)	279	-do-	Anonymous (1982)
Mandya(Karnataka)	400	Red sandy loam	Hunsigi (1981)
Kolhapur(Maharashtra)	335	-	Mane & Salunkhe (1978)

and Misra 1956); 136 kg at Mandya (Hunsigi, 1981), 119.5 kg at Lucknow (Jafri *et al*, 1978) and 119.2 kg at Jalandhar (Rattan and Singh, 1954).

Why more nitrogen to ratoons?

Higher dose of nitrogen for ratoons than the corresponding plant crops may be due to (a) depletion of N content in rhizospheric soil, (b) microbial tie up of soil available N and (c) lesser nitrate reductase activity of ratoon plants. Mohan Rao *et al.* (1956) at Anakapalle and Jafri and Alam (1982) at Lucknow analysed the N content in rhizospheric soils at the time of harvesting of plant cane and found decrease in level of available nitrogen. This depleted level of nitrogen may be partly responsible for increased level of nitrogen application need for succeeding ratoons.

Initially sugarcane ratoons start functioning on old stubble roots. After 3-6 weeks, old root system ceases to function and new root system is formed. Thus, new root system of ratoon crop is surrounded by old decomposing roots of high C:N ratio (Table 7.2). Microbial activity is likely to be

Table 7.2 : Composition of old sugarcane roots (oven dry basis)

Constituent	Concentration
Organic matter (%)	91.46
Ash (%)	8.54
Total carbon (%)	53.05
N (%)	0.51
P (%)	0.09
K (%)	0.80
Na (%)	0.38
Ca (%)	0.20
Fe (ppm)	2083.1
Mn (ppm)	276.8
Cu (ppm)	17.44
Zn (ppm)	26.15

Source : Singh and Agrawal, 1981

high in the root zone as a result of the decomposing old root system which immobilises a part of applied N. In a study conducted at the Indian Institute of Sugarcane Research, Lucknow by Singh and Agrawal (1981), it was observed that about 3.9 t/ha of sugarcane roots are added at the time of ratooning having a C:N ratio of 10:6 with N content as 0.51 per cent. About 35-40 kg N will be required to lower C:N ratio to 30:1 and to raise N content to 1.76 per cent. This added dose will stimulate the root decomposition and nitrogen mineralization. Therefore, higher rates of N application may be required to overcome microbial tie up.

It has also been observed that for the induction of the same amount of nitrate reductase activity *in vivo* a ratoon crop requires relatively more NO_3^- nitrogen than the corresponding plant crop (Rai *et al*., 1989).

Nitrogen use efficiency of ratoon crop in comparison to plant cane : Lakshmikantham (1973) on the basis of his studies at Anakapalle (Andhra Pradesh) concluded that ratoons are less efficient utilizers of nitrogen as compared to plant canes. In support of his statement, he calculated that cane production per kg N per hectare were 1.03 and 0.91 tonnes in plant crop and 0.85 and/or 0.80 tonnes in first ratoon of the variety Co 419 and Co 997, respectively. This statement is based on the total production of cane at 112 kg N per hectare (Table 7.3). Misra and Mathur (1983) had a closer look at the same data and found out the fallacy in the conclusion drawn on nitrogen use efficiency by ratoon crop. They calculated the increase in cane yield per kg N per hectare by using figures for excess production of cane per hectare over 0 kg N dose (Table 7.4). It clearly

Table 7.3: Nitrogen use efficiency (Cane yield t/ha/kg N) of plant and ratoon crops of sugarcane calculated on the basis of yield obtained at different levels of N application

Variety	N(Kg/ha)	Cane yield(t/ha)		N use efficiency	
		Plant cane	Ratoon	Plant cane	Ratoon
Co 419	0	57.67	46.58	—	—
	112	115.03	94.98	1.03	0.85
	224	120.09	113.07	0.54	0.50
Co. 997	0	64.64	50.51	—	—
	112	101.39	89.85	0.91	0.80
	224	105.46	97.63	0.47	0.44

Source: Lakshmikantham, 1973

Table 7.4: Nitrogen use efficiency (Cane yield t/ha/kg N) of plant and ratoon crops of sugarcane calculated on the basis of excess cane yield (t/ha) obtained over N dose

Variety	N (kg/ha)	Cane yield (t/ha) Plant cane	Cane yield (t/ha) Ratoon	N use efficiency Plant cane	N use efficiency Ratoon
Co 419	112	58.36	48.40	0.52	0.43
	224	62.42	67.49	0.28	0.30
Co 997	112	36.75	39.34	0.33	0.35
	224	40.82	47.12	0.18	0.21

Source :Misra and Mathur, 1983

shows that at 224 kg N per hectare, the net excess production from ratoon was 67.49 and 47.12 tonnes of cane as compared to only 62.42 and 40.82 tonnes in plant cane in the two varieties i.e. CO 419 and Co 997, respectively. At 112 kg N dose, the difference was not so clearly marked, particularly in variety Co 419 in favour of ratoon crop and that is quite obvious as with this low dose, which is lower by about half of the recommended dose for plant crop at Anakapalle, the already depleted soil in case of ratoon fields could not sustain higher production of ratoon cane. However, when the dose of nitrogen was raised to the required recommended level of 224 kg N per hectare, the ratoon crop derived better advantage from it. It is important to note that contrary to Lakshmikantham's contention, at proper dose of nitrogen ratoon is better utilizer of applied nitrogen than the plant crop.

7.2 **Response to phosphorus :** In a large number of trials, applied phosphate did not influence yield or quality of sugarcane ratoon to any appreciable extent. It was because of the fact that in most of the cases, the soils were having high level of this nutrient. However, the need for phosphate application ranging from 30 to 100 kg P_2O_5 per hectare has been stressed in some experiments (Table 7.5). In India, cane response to applied phosphorus is limited in certain soils like red/mixed, red and black soils of Karnataka, calcareous soils of Bihar and at some pockets in alluvial soils of Uttar Pradesh,

Table 7.5 : Effect of phosphorus application on cane yield and juice quality of sugarcane ratoon at different locations of India

Location	Significant yield increase up to P_2O_5 (Kg/ha)	Soil type	Juice quality	Reference
		Subtropical India		
Lucknow (UP)	NR	Alluvial, Sandy loam	NA	Jafri & Alam (1980)
Jalandhar (Punjab)	30	-do-	NA	Kapur & Kanwar (1991)
Kheri (Punjab)	60	Sandy clay loam	NA	Kapur & Kanwar (1991)
Jagadhari and Ladwa (Haryana)	50	In certain soils Alluvial, Sandy loam	NA	Singh *et al.* (1977)
New Delhi	80	-do-	NA	Prasad & Parashar (1989)
Pusa (Bihar)	50	Calcareous alluvium	NA	Thakur *et al.* (1979)
		Tropical India		
Anakapalle (AP)	NR	Sandy loam to clay loam	NA	Mohan Rao *et al.* (1956)
Anakapalle (AP)	NR	-do-	NA	Lakshmikantham (1973)
Mandya (Karnataka)	100	Red sandy loam	-	Hunsigi (1981)
Nellikuppam (Tamil Nadu)	56	Coastal alluvium, Sandy loam	NA	Anonymous (1982)
Godavari (AP)	75	—	I	Vaidya (1965)

NR–No response, NA–Not affected, I–Improved

Haryana and Punjab. But responses are absent in medium to heavy black soils of Karnataka, Maharashtra and Madhya Pradesh. Application of phosphorus also did not influence either the cane yield or juice quality of ratoon at Anakapalle (Mohan Rao *et al.*, 1956) and in East Godavari district (Rege and Patwardhan, 1945) of Andhra Pradesh.

In general, ratoon crop of sugarcane is relatively more responsive to phosphate application than plant cane as observed at Mandya, Karnataka (Hunsigi, 1981), in certain pockets of Ladwa and Jagadhari in Haryana (Singh *et al.*,1977) and at Jalandhar and Kheri in Punjab (Kapur and Kanwar, 1991). Ratoon yield response to per kg P_2O_5 per hectare recorded were 263 kg at Jalandhar and 243 kg at Kheri in Punjab, 153 kg at Pusa (Bihar), 113 kg at New Delhi and 75 kg at Mandya (Karnataka).

7.3 **Response to potassium :** Sugarcane crop requires an excess amount of potassium for its growth and development. Since plant crop uptakes a considerable amount of available potassium from soil during its life cycle, the ratoon crop has to start at a comparatively lower initial supply (Jafri and Alam, 1982). However, field experiments conducted in India so far have indicated that sugarcane crop does not respond to applied potassium (Table 7.6) except in some localised areas of Andhra Pradesh, Karnataka, Maharashtra, Orissa and Assam (Srivastava and Seth, 1990). Studies in red sandy loam soil of Mandya (Karnataka) indicated a consistent response to applied potassium and responses varying from 60-170 kg/ha/kg of K_2O were obtained at 100kg of K_2O /ha level of application (Hunsigi, 1981).

The lack of response of sugarcane to applied potassium in alluvial soils of *Indo-Gangatic* plains is ascribed to sufficient

Table 7.6 : Effect of potassium application on cane yield and juice quality of sugarcane ratoon at different locations of India

Location	Significant yield increase up to K_2O (Kg/ha)	Soil Type	Juice Quality	Reference
Subtropical India				
Lucknow (UP)	NR	Alluvial sandy loam	NA	Anonymous (1959, 1960, 1961)
Lucknow (UP)	NR	-do-	NA	Jafri and Alam (1980)
Kanpur (UP)	NR	-do-	I	Rathi & Singh (1983)
Jalandhar (Punjab)	NR	Alluvial sandy loam	I	Kapur & Kanwar (1991)
Kheri (Punjab)	NR	-do-	I	Kapur & Kanwar (1991)
Jagadhari (Haryana)	NR	-do-	NA	Singh *et al.*(1977)
Tropical India				
West Godavari (AP)	NR	-	-	Vaidya (1965)
Mandya (Karnataka)	100	Red sandy loam	I	Hunsigi (1981)
Nellikuppam (Tamil Nadu)	112	Sandy loam	-	Anonymous (1982)

NR–No response, NA–Not affected, I–Improved

K reserves due to illite clay mineral in these soils (Kanwar, 1976).

Potassium and juice quality : Effect of potassium in improving cane quality is debatable (Table 7.6). However, in a critical appraisal, potassium application seems to improve the juice quality of ratoon more than the plant cane. Despite contradictions on the role of potassium in improving cane quality, the consensus is that addition of K improves sucrose per cent cane by increasing extraction percentage and by reducing the fibre content. This has more relevance in ratoons as they tend to contain more fibre than the plant crop (Hunsigi, 1981). Fritz (1974) has offered physiological explanation. for improved juice quality following potassium application. He observed that the quality is improved only when K application reduces leaf N coupled with a reduction in tissue moisture content. Thus, adequate K is required to utilize the unassimilated nitrogen in the cane to bring about the stage of maturity when reducing sugars are converted to recoverable sugars.

7.4 **Response to micro-nutrients :** Sometimes the deficiency of minor nutrient play a major role in limiting the growth of the crop. In severe deficient conditions, plants show some abnormal symptoms. But under less pronounced deficiency, the crop appears healthy but such crop may also respond well if the deficiencies are made good.

Iron chlorosis : In certain fields, deficiency symptoms (Fig. 7.1) of iron are generally seen, more so in ratoon. Symptoms of iron deficiency on the plants are characterised by intervenal chlorosis (Fig. 7.2) because plants deprived of an adequate supply of iron fail to synthesize adequate chlorophyll and thus become chlorotic. This malady is also known as lime-induced chlorosis, because it is primarily a potential problem on calcareous soils. Since iron is immobile in plant body so the symptoms of iron deficiency start appearing on younger leaves. In general, the initial symptoms of chlorosis appear with the onset of hot weather and increase thereafter (Agrawal and Lal, 1984). In certain cases, the affected plants become chlorotic in appearance as soon as they emerge from the soil. The persistence of this disorder

Fig. 7.1 : Sugarcane crop affected with iron chlorosis

for about four months, particularly in early phases of crop growth, may lead to 74% loss in cane weight, 41.7% loss in sucrose content, 73.3% increase in reducing sugars and 48.7% reduction in recovery (Yadav and Yaduvanshi, 1989). In extreme conditions younger leaves become completely white which may depress the growth severely and plants may die. Agrawal and Lal (1987) reported that in

Fig. 7.2 : Iron deficient sugarcane plant (left) and healthy plant (right)

vertisols of Madhya Pradesh, containing $CaCO_3$ as high as 25 per cent, sometimes the lime-induced chlorosis resulted in a complete failure of the cane crop.

There is wide diversity in opinions with regard to the mechanism and cause of lime-induced chlorosis. However, excess quantities of calcium in such soils, resulting in inactivation of iron in the soil as well as in plants, are held responsible for iron chlorosis. Since iron is probably largely assimilated by plants as ferrous iron, calcareous soils would not favour iron availability because the excess lime favours oxidation of iron to the less active ferric state and also fixation in compounds of low biological activity (Brown, 1956). The effect of excess calcium in a soil is also generally associated with high pH and consequently unavailability of iron (De Kock, 1955). It is, therefore, difficult to separate this effect of calcium from the effect of soil pH on iron availability.

The remedial measure consists of spraying of 1-2% ferrous sulphate ($FeSO_4$) solution in water twice at fortnightly interval (Prasada Rao, 1980; Hunsigi, 1982). Application of pyrite in soil (0.5 t/ha) was also found effective (Table 7.7) for

Table 7.7 : Effect of soil application of pyrite (FeS_2) in improving yield and quality of sugarcane ratoon by correcting lime-induced iron chlorosis on vertisols of Mandsaur area of Madhya Pradesh (India)

Treatment	Cane yield (t/ha)	Juice quality	
		Brix	Pol%
Control	38.1	16.76	14.28
Pyrite(0.5 t/ha)	71.9	17.45	14.71
CD (0.05)	13.8	0.36	0.38

Source :Agrawal and Lal, 1987

correcting this nutritional disorder and improving yield and quality of sugarcane ratoons (Agrawal and Lal, 1987).

Other micro-nutrients : Sugarcane seldom exhibits micro-nutrient deficiency, other than chlorosis, under field conditions (Zende, 1973). Therefore, the deficiencies may be diagnosed with the help of foliar or soil analysis. Yield of sugarcane may be improved by the application of micro-nutrients, if deficient, either in soil or on foliage. Significant yield response to foliar application of Zn, Fe, Mn and Cu has been recorded (Yadav and Yaduvanshi, 1987) at Lucknow (Table 7.8). However, micro-nutrients should be applied in appropriate doses, only when deficiency of particular nutrient is confirmed. Blanket use of micro-nutrients should be avoided.

Table 7.8 :Effect of foliar application of micronutrients (0.5% solution) on cane and sugar yields of ratoon at Lucknow

Spray application	Yield (t/ha)	
	Cane	Sugar
Zn as $ZnSO_4 . 7H_2O$	50.2	6.02
Fe as $FeSO_4 . 7H_2O$	49.3	6.00
Mn as $MnCl_2 . 4H_2O$	44.2	5.21
Cu as $CuSO_4 . 5H_2O$	43.8	5.03
Control (without spray)	35.5	4.07
Control (with water spray)	37.7	4.34
CD at 1%	5.2	-

Note : The DTPA extractable (Lindsay and Norvell, 1978) Fe, Cu, Mn, and Zn in the soil of experimental fields were 74.87, 6.19, 13.89 and 0.82 ppm, respectively.

Source :Yadav *et al.*, 1987

7.5 **Limits of deficiency and sufficiency of nutrients in soil and plants of ratoon crop of sugarcane :** Wide variation in fertility of different soils are recognized by soil testing. Thus, generalized blanket fertilizer recommendations for sugarcane production in different groups of soils may not be economical. The level of nutrients to be maintained in soil and index tissues of plants at specified ages of the crop have been determined for realizing high cane yields. The limits of deficiency and sufficiency of major nutrients (NPK) in soil and plants of ratoon crop of sugarcane during formative phase as observed at Lucknow (Subtropical India) are presented in Table 7.9. It is evident from the table that due to variation in absorbing capacity, deficiency and sufficiency limits of nutrients in index tissue differ from variety to variety though the soil limits remain the same.

Table 7.9 : Limits of deficiency and sufficiency of nutrients in soil and plants during formative phase of sugarcane at Lucknow (Subtropical India)

Soil limits (Kg/ha)	Plant limits (%) in Co 1158	Co 1148
Alkaline hydrolysable N	**N content in 3-6 leaf blades**	
<200 (low)	<1.541	<1.495
>400 (High)	>1.732	>2.283
Olsen's P (P_2O_5)	**P content in 3-6 leaf blades**	
<25 (low)	< 0.088	<0.066
>50 (High)	>0.106	>0.086
Exchangeable K (K_2O)	**K content in 3-6 leaf sheath**	
<100 (low)	<1.921	<2.500
>250 (High)	>2.032	>3.050

Source : Jafri, 1983

In tropical part of India, 1.8% nitrogen in leaf laminae, 0.08% phosphorus and 1.99% potassium in leaf sheath of 3-6 leaves should be maintained during grand growth phase for realizing satisfactory cane yields (Prasada Rao *et al.*,1987a).

No critical limits for soils or in plant index tissues with respect to micro-nutrients for sugarcane have been reported so far in India. However, the levels of micro-nutrients in the index tissues below which deficiency symptoms are likely to occur were reported for manganese (3 ppm), iron (10 ppm) zinc (10 ppm) and copper (3.5 ppm) from other countries (Prasada Rao *et al.*, 1987a)

Relationship between available soil nutrients and their concentrations in plant tissue : Relationship between available soil nutrients and their concentration in plant index tissues are shown in Table 7.10. Most of the

Table 7.10 : Correlation between available soil nutrients and their concentration in plant tissues

Nutrient (kg/ha) in soil (x)	Nutrient (%) in index tissues (Y)	'r'	Regression equation
	Variety Co 1148		
Available N (alk.$KMnO_4$ method)	Nitrogen	0.82**	Y=0.569+0.00876x
Olsen's P (P_2O_5)	Phosphorus	0.77**	Y=0.450+0.0002426 x
Exchangeable K (K_2O)	Potassium	0.82**	Y=1.530+0.0036 x
	Variety Co 1158		
Available N (alk.$KMnO_4$) method	Nitrogen	0.68**	Y=0.8815+0.0033 x
Olsen's P (P_2O_5)	Phosphorus	0.70**	Y=0.0647+0.0003 x
Exchangeable K (K_2O)	Potassium	0.85**	Y=0.8460+0.0066 x

Source :Jafri, 1979 ,** Significant

correlation coefficients are significant. Regression curves depicting increase in K in plant index tissues with the increase of soil exchangeable K have also been shown in Fig. 7.3. The behaviour of two varieties in accumulating K in plant tissues differs widely. Nutrient contents and index tissues are also correlated with cane yield (Table 7.11).

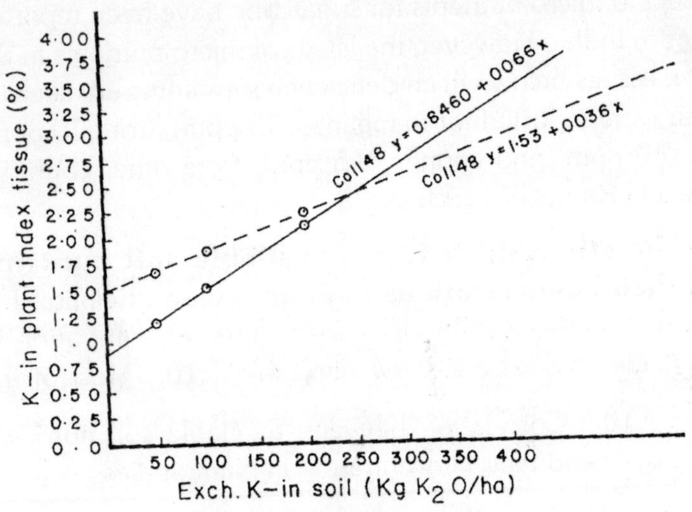

FIg. 7.3 : Correlation between soil exchangeable K (kg K_2O/ha) and K in palnt index tissue (%) in ratoon crop of two varieties (Co 1158 and Co 1148)

Table 7.11. Correlation between nutrient content in index tissues and yield of ratoon cane (Cv. Co 1148)

Nutrient (kg/ha) in soil (x)	Nutrient (%) in index tissues (y)	'r'	Regression equation
Nitrogen (kg/ha)	Cane yield	0.93**	Y=2.180+ 54.260
Phosphorus	Cane yield	0.73**	Y=21.057 + 1198.1947
Potassium	Cane yield	0.79**	Y=-27.862 + 49.9071

Source: Jafri, 1979 ** Significant

Fertilizer requirement for different targeted yields : Prasad *et al.* (1984) suggested following equation (Table 7.12) based on soil test values to work out the fertilizer requirement of sugarcane ratoon for different yield targets in calcareous soils of Bihar. On the basis of these equations they calculated the N, P_2O_5 and K_2O requirement (kg/ha) of sugarcane ratoon for different yield targets (Table 7.13) in soils with varying soil test values.

7.6 **Residual effects of fertilizer and manures** : Sharma *et al.*, (1981) at New Delhi, noted a significant carry-over effect of nitrogen applied to plant crop in succeeding ratoon. The carry over effects were more when N was applied as slow-release N-fertilizer, sulphur coated urea or nitrification

Table 7.12 . Equations to work out fertilizer requirement of sugarcane ratoon crop for different targeted yields

Nutrients	Targeted yield equation	NR (kg/q) of cane	CS%	CF%	Range of soil test values (kg/ha)
F N	1.01T-1.65 SN	0.211	33.4	20.9	197-394 (alk $KMnO_4$)
F P_2O_5	0.24T-3.00 SN	0.024	33.4	10.0	14.4-75.6 (Olsen's P)
F K_2O	0.11T-0.118 SK	0.156	17.0	144.0	101-185 (ammo.acet.)

NR = Nutrient requirement (kg/q ratoon sugar cane)

CS = Contribution from soil as percentage of its available nutrient

CF = Contribution from fertilizer as percentage of its nutrient content

F, T and S are Fertilizer (kg/ha), Targets (q/ha) and Available nutrients in soil (kg/ha), respectively.

Source : Prasad *et al.*,1984

Table 7.13 : Nitrogen, phosphorus and potash requirement (kg/ha) of sugarcane ratoon for different targeted yields in soil with varying soil test values

Soil test values (kg/ha)	Yield target (q/ha) 500	700
Available N	**N required**	
200	175	377
250	93	295
300	10	212
Available P_2O_5	**P_2O_5 required**	
20	59	109
30	29	77
40	-	46
Available K_2O	**K_2O required**	
100	43	65
150	37	59
200	31	53

Source :Prasad *et al.*,1984

inhibitor treated urea or urea coated with *neem* cake to the plant crop. Yadav *et al.* (1990) did not observe appreciable residual influence of various slow released nitrogenous fertilizer or nitrification inhibitors on the yield of subsequent ratoon crop at Lucknow. In most of the other studies too residual carry over of nitrogen applied through chemical fertilizers was negligible (Srivastava, 1988). Adequate direct application of N is, therefore, suggested for increasing the yield of ratoon crop. It is well known that the N fertilizers are lost through leaching and volatilization and one should not expect much of residual effect of these fertilizers. However, bulky organics applied to plant crop for N supply, in general, show better residual effect in ratoon crop (Table 7.14).

Some people are of the opinion that the ratoon crop yield more if raised from plant crop with comparatively heavier doses of nitrogen. Mathur *et al.* (1968) at Gorakhpur got yield of subsequent ratoon cane higher from plots receiving heavier doses ranging from 269 to 300 kg N/ha in plant crop as compared to those receiving lower doses of 135 to 150 kg N/ha in plant cane, even though the dose for ratoon remained the same (75 kg N/ha) in each case of ratoon. Yadav and Sharma (1982) also reported that an increase in the application of N up to 225 kg/ha to the plant crop increased the yield of the subsequent ratoon crop significantly. On the other hand ratoons of plant crops which received 100 and 200 kg N/ha, did not differ in yield at Lucknow (Anonymous, 1958). Similarly Vaidya and Kulkarni (1962) also observed

Table 7.14 : Residual effect of chemical fertilizer and organic manure (applied to plant crop) on succeeding ratoon yield at Mazaffarnagar (U.P.)

Manurial treatment to plant cane	Ratoon yield (t/ha)
Control	35.4
Ammonium sulphate (60 kg N/ha)	36.2
Ammonium sulphate (120 kg N/ha)	40.6
FYM (60 kg N/ha)	38.0
FYM (120 kg N/ha)	42.1
FYM + Ammonium sulphate (30+30 kg N/ha)	37.4
FYM + Ammonium sulphate (60+60 kg N/ha)	44.2
CD (0.05)	6.4

Source :Ann. Rep. Sugarcane Research in U.P. during 1940-41, p. 87

no residual effect of higher nitrogen doses applied to previous crop on the subsequent ratoon in Maharashtra. Moreover, it seems unwise to fertilize plant crop with high doses, and for ratoon to rely upon the residual effects. A dose of fertilizer given in time even in lower quantities would be more effective than the assumed residues.

Phosphatic fertilizers are usually applied as single superphosphate or diammonium phosphate in sugarcane. The soluble phosphate ions in these materials are absorbed by plant roots and 10-20 per cent of applied phosphate is utilized by current sugarcane crop. The remaining is partly fixed by the soil and is partly shared in forming the definite reaction products of phosphates. This residual phosphate is available for a longer period depending upon the soil characteristics. Jafri and Alam (1982) recorded significant increase in level of available phosphorus in soil of ratoon field due to application of P fertilizer in preceding plant crop at Lucknow. Thus, an application of P to ratoon field which had received dressing of P in the plant crop stage yielded no response. Similarly, Kapur and Kanwar (1991) recorded residual effect of P on ratoon crop in Punjab and stated that if the plant crop was supplied with 60 kg P_2O_5/ha in Jalandhar, and 90 kg/ha in Kheri (Sangrur) soils, then ratoon crop did not require any additional P fertilizer. Otherwise application of 30 and 60 kg of P_2O_5/ha is needed for ratoon grown in Jalandhar and Kheri soils, respectively (Table 7.15)

Types of fertilizers

There are several nitrogenous (ammonium sulphate, ammonium sulphate nitrate, calcium ammonium nitrate, ammonium choloride and urea), phosphatic (single superphosphate and rock phosphate) and potassic (Potassium

Table 7.15 : Direct and residual effects of phosphatic and potassium fertilizers on ratoon cane (Cv CoJ 64) yield (t/ha) at Jalandhar and Kheri (Punjab)

Nutrient	Jalandhar		Kheri (Sangrur)	
(kg/ha)	Direct	Residual	Direct	Residual
P_2O_5				
0	61.3	60.8	53.8	52.8
30	69.2	64.1	61.9	55.5
60	71.9	68.9	68.4	63.3
90	72.4	70.8	70.0	68.6
CD (0.05)	-	5.8	-	6.2
K_2O				
0	69.9	66.6	63.4	60.9
30	71.7	66.7	64.5	60.6
60	71.9	68.9	66.4	62.3
90	70.9	67.7	65.7	61.3
CD (0.05)	-	NS	-	NS

Place	Soil type	Available nutrient (mg/kg soil)			Soil pH
		N	P	K	
Jalandhar	Sandy loam	122	6.2	49	7.9
Kheri	Sandy clay loam	110	5.9	58	8.4

Source : Kapur and Kanwar, 1991

chloride and potassium sulphate) fertilizers being used in India. It also includes fertilizers such as monoammonium phosphate and diammonium phosphate etc. In trials, using different carriers of N in the form of nitrate, amide or ammonium no worthwhile differences attributable to the form and the carrier has been noted (Srivastava and Seth, 1990). Anhydrous ammonia was also found suitable for ratoon, but the Indian implements are not suitable for its application because anhydrous ammonia is to be injected deep into the soil followed by immediate covering of the furrows to avoid volatilization loss of the gas. The efficiency of various forms of phosphatic fertilizers is associated with the chemical characteristics of the soil being fertilized. Water soluble P carriers like superphosphate, mono-ammonium phosphate and diammonium phosphate are suitable in most of the situations. On acid soils, rock phosphate and monocalcium phosphate are equally effective (Srivastava and Seth, 1990). No appreciable difference has been noted between the potassium chloride and potassium sulphate as potassium supplier to sugarcane. Recently, an indigenous K source i.e. potassium scheonite (24% K_2O) was compared with muriate of potash for ratoon production at Mandya (Karnataka). It proved equally effective (Hunsigi, 1982)

The choice of the particular nitrogen and potassic fertilizer should be based on their cost per unit weight of the nutrient and for phosphatic fertilizers their relative efficiency in given soil is to be considered.

The use of nutrients through organic manures has a special significance for ratoon crop. A proper combination of both organic manures and inorganic fertilizers has been considered desirable for long term maintenance of cane and sugar yields and to obtain better response from applied fertilizers. Organic manures also increase ratooning capacity of sugarcane. Studies conducted in Muzaffarnagar and Saharanpur districts of western Uttar Pradesh indicated that one part of N through organic manure and two parts of N through inorganic fertilizers had a better effect in increasing ratoon yields and juice quality (Agrawal, 1981).

7.8 **Methods of fertilizer application**

Adoption of proper method of fertilizer application is essential to minimize the nutrient losses from the soil and for increasing fertilizer use efficiency. Fertilizers should be placed close to the root zone to enable the roots to derive its benefit immediately. Besides increasing cane yield (Table 7.16) proper placement also reduces volatilization losses of nitrogenous fertilizers and lowers the fixation of phosphatic fertilizers. In practice, this may be done by making 8-10 cm deep furrows on either side of cane rows by suitable implements and then covering it after placing the fertilizers in them.

Table 7.16 : Effect of method of nitrogenous fertilizer application on yield (t/ha) of sugarcane ratoon at Lucknow.

Application method	N level (kg/ha)				Mean
	0	75	150	225	
Band placement	-	59.9	69.0	72.9	67.3
Broadcasting	-	49.6	55.0	57.0	54.1
Mean	-	40.2	54.8	62.0	

Factor	CD (0.05)
Application method (M)	4.0
N levels except control	4.9
MxN	6.9
Control *vs* other N levels	8.5

Source: Bhattacharya *et al.*, 1984

7.9 Time of fertilizer application

Time of fertilizer application assumes great significance in deriving its benefits. If the dose of N to be applied is 150 kg/ha or less its application in a single dose at ratoon initiation would be the best practice. Alternatively, the nitrogenous fertilizers may be applied in two or three splits (Table 7.17). Early application of nitrogen to ratoon fields is necessary to ensure that new growth is not starved of N through competition with soil micro-organisms. With the mass of old roots and the underground stubbles decomposing under the influence of micro-organisms, an adequate amount of available soil N is immobilized. Consequently, utilization of nitrogen by the micro-organisms depletes the nitrogen leaving little for the newly developing root system. Thus, even in case of split application, one-third to half dose of nitrogen should be applied immediately after harvesting of the preceding crop at the time of first irrigation. The balance amount of N, in one or two equal doses, should be given within 90-100 days or in one lot at the onset of *monsoon*. It is not desirable to postpone application of entire dose of N to the time of break of rains as it delays maturity and lowers the cane juice quality.

The timing of nitrogen fertilization should also be adjusted to the anticipated availability of irrigation or rainfall, since yields are expected to be reduced due to lack of irrigation water. Fertilizer rates also need to be cut down in proportion to the anticipated yield potential. However, as an insurance dose, 25 to 30% of the estimated fertilizer should be applied at ratoon initiation. The remaining dose may be given in one or two splits coinciding with irrigation available or in one lot with the onset of *monsoon* (Srivastava and Seth, 1990).

Full dose of phosphorus and potassium, where necessary, should be given alongwith the first dose of nitrogen

Table 7.17: Proper time of nitrogen application to sugarcane ratoon at different places of India

Location	N dose (kg/ha)	Soil type	Proper time of N application	Reference
Lucknow (UP)	200	Sandy loam	1/3 at IR + 1/3 in April + 1/3 in May	Singh and Sharma (1973)
Seorahi (UP)	180	*Bhat* soils	1/3 at IR+ 1/3 in May + 1/3 in June	Chaudhary *et al.* (1992)
Gorakhpur(UP)	120	Sandy loam	Full at IR or 1/2 at IR + ½ at the onset of *monsoon*	Tandon *et al.* (1957)
Shahjahanpur (UP)	120	Sandy loam	-do-	Tandon *et al.* (1956)
Muzaffarnagar (UP)	120	Sandy loam	-do-	Tandon *et al.* (1957)
Jalandhar (Punjab)	225	Sandy loam	1/3 in March+ 1/3 in April + 1/3 in May	Kanwar (1980)
Jagadhari (Haryana)	225	Sandy loam	1/3 in Feb-Mar+ 1/3 in April-May + 1/3 in May-June	Singh *et al.*(1977)
Pusa	150	Sandy loam	Full at IR or 3/4 at IR + 1/4 at earthing	Anonymous (1957)
Mandya (Karnataka)	400	Red sandy loam	1/3 at IR + 1/3 (8WAIR)+ 1/3 (14WAIR)	Hunsigi (1982)
Anakapalle (AP)	224	Clay loam	1/2 at IR + 1/2 at tillering	Anonymous (1972)

IR = Initiation of ratoon, WAIR = Week after initiation of ratoon at ratoon initiation (Kanwar 1980 and Adhate, 1960).

7.10 **Manuring in relation to irrigation**

There is a high positive interaction between nitrogen and irrigation (Mathur and Haider, 1940). Heavy dressing of nitrogen without adequate number of irrigations proved less effective and *vice versa*. Under deficient irrigation conditions, the application of even one irrigation to ratoon crop produces distinctly higher yields, provided early application of N is also made (Nath and Mathur, 1964; Mathur and Singh, 1968). Growing sugarcane ratoon without irrigation and applying full dose of N at the onset of *monsoon* is not desirable at all (Table 7.18).

Table 7.18 : Effect of time of N application on cane juice quality of ratoon sugarcane at Shahjahanpur (Uttar Pradesh) (mean of data from 1952-53, 1953-54 and 1954-55)

Time of application	Juice quality in December		
	Brix	Pol %	Purity %
A . Full at ratoon initiation	18.81	16.53	86.94
B . Full at the break of rains	18.25	15.75	86.00
C . Half as 'A' and half as 'B'	18.53	16.07	86.71
D . Control (no nitrogen)	18.48	16.09	87.09

Note : N dose was 134 kg/ha

Source: Misra, 1966

7.11 **Effect of fertilizer practices of plant cane on ratoon yield**

Bhoj (1976) suggested that plant crop should be adequately manured, if good ratoon is aimed at. Jafri and Alam (1982) stated that the absence of fertilizer in plant cane but its application to succeeding ratoon produced significantly low yield indicating that the ratoon performs well only when the plant crop has been fertilized sufficiently. Adequate manuring and irrigation of plant cane provide uniformly good plant population which is reported to be an important factor in ratooning (Raheja and Misra, 1954; Bhoj and Mathur, 1965)

8. Water Management

Water is an essential input in crop production. Sugarcane has a peculiar situation so far as water management is concerned. The crop faces drought conditions during summer months in formative phase and waterlogging during *monsoon* in grand growth phase. In general, situations of water availability to sugarcane crop may be deficient, optimum or excessive. Under all the situations water is to be managed so judiciously that maximum benefit is obtained with minimum harm to plant and soil. For efficient water management, a correct understanding of the relationship between climatic conditions, physicochemical properties of the soil, quality of water, soil depth, depth of water table effective root zone depth, intake rate of water in soil and crop stage is essential to make rational decisions concerning time, method and amount of water application through irrigations during non-rainy season and disposal of excess water, if any, during rainy season. Thus, water management technology will vary from region to region due to variations in climatic conditions, soil and cultural practices.

8.1. Water requirement (WR)

It is the total amount of water needed for raising a crop successfully. In case of sugarcane, it includes the amount of water for meeting needs of evaporation, transpiration and metabolic activities (known as consumptive use), losses during application of water and water needed for land preparation as pre-planting irrigation. Under field conditions, WR is met by effective rainfall, contribution from ground water table (if it is within the reach of the root system) and irrigations.

Due to longer crop duration, WR of sugarcane is quite high and also varies greatly from place to place (Table 8.1) depending upon the climatic conditions of the area, soil type and crop duration. WR of sugarcane ranged from 1400 mm in Bihar to 3000 mm in Maharashtra. The WR of *adsali* crop of sugarcane in Maharashtra is 3500 mm. It is estimated that to produce one tonne of cane 200-250 tonnes of water is needed.

Determining WR of sugarcane: Leather (1910) took up studies related to WR of sugarcane first time in India. Since then considerable work on WR of this crop has been done at different research centres in India. To determine WR under field conditions the following items are to be considered:

- Consumptive use or evapotranspiration
- Irrigation losses, including seepage, run-off and deep percolation, expressed as irrigation efficiencies.

Consumptive use is estimated as evapotranspiration values. Under cropped conditions, evaporation and transpiration occur simultaneously and it is difficult to separate these two losses. The two processes are, therefore, combined into one term evapotranspiration'(ET). The third part of consumptive use of water i.e., water used in plant metabolic activities, is hardly 1% of the total ET so this water loss is considered as negligible.

ET values can be computed by various empirical formulae especially those of Blaney-Criddle(1945), Thornthwaite(1948), and modified Penman(1948). Almost all the empirical formulae need a huge data on temperature, humidity, wind velocity, sunshine hours etc, to estimate the evapotranspiration. These data are either not available or partly available for the command area of any irrigation project.

It is also not reliable to estimate ET from a few years' meteorological data. It has been found that open pan evaporimeters gave the best correlation as compared to these formulae for the estimation of ET. Evaporation pans provide a measurement of integrated effect of radiation, temperature and relative humidity on evaporation from specific open water surface. Pan evaporimeters are considered more accurate in estimating short term fluctuations in ET than empirical formulae. Moreover, the pan evaporation data are easy to collect, simple, less laborious, less expensive and can be collected in short time. The Standard U.S. Bureau Class-A Pan Evaporimeter is widely used for measurement of evaporation.

The crop evapotranspiration (ETc) values can be estimated as:

ETc = ET x Kc

E T = PE x K pan

Where, ETc = Crop evapotranspiration (mm/day)

PE = Pan evaporation (mm/day)

K pan = Pan coefficient generally 0.7

or 0.8 depending upon location of pan

Kc = Crop coefficient

The crop coefficient (Kc) varies according to crop characteristics, stage and rate of growth. It is generally 0.6 during germination and settling stage, 0.8 during early growth stage, 0.8 to 1.0 during tillering and grand growth stages and 0.8 during maturity stage of sugarcane crop.

Table 8.1: Water requirement (WR) of sugarcane in different States of India

State	WR (ha cm)
Subtropical India	
Bihar	140
Uttar Pradesh	160-180
Punjab	170-180
Tropical India	
Andhra Pradesh	160-170
Tamil Nadu	180
Karnataka	200-240
Maharashtra	
Plant cane (Seasonal)	250
Plant cane (Pre-seasonal)	300
Plant cane (*Adsali*)	350
Ratoon	300
Madhya Pradesh	270

Factors affecting WR of sugarcane

Any factor which influences evapotranspiration and /or direct water losses during its application affects the WR of sugarcane.

A. Factors affecting ET

ET depends on a number of interacting factors such as climatic conditions, nature of soil, stage of crop growth, canopy structure and length of growing season.

i) Climate: The ET from a cropped field primarily depends upon climatic parameters. ET involves conversion of moisture from soil and plants into water vapours that needs energy of 540 cal. per gram of water which is provided by solar radiation. It is a physical phenomenon governed by the incident energy input and sink strength of the wind at a place. Role of wind is to carry evaporated and transpired water vapour away from leaves and soil surface thus steeping the gradients (sink strength). If there is active green vegetation covering the ground completely under non-limiting water supply, the quantity of water lost by vegetation per day is independent of its species. Thus, water lost by a crop is not a physiological process.

ii) Soil: Water need for ET of a crop growing under the same climatic conditions but on different soil will remain the same. Soil acts as a reservoir of water for plants to draw upon so it governs only the depth and frequency factors in irrigation. Under the condition of low soil water holding capacity (WHC) and shallow soil depth, irrigation depth is smaller while on deep soil and with high WHC the irrigation depth is relatively greater. Lower WHC and soil depth increases number of irrigations. Greater the number of irrigations, higher the application losses that increases the WR of crop.

iii) Crop Stage: ET/day in sugarcane differs as per its growth stage. It increases from germination to full canopy stage and continues up to the end of grand growth phase and then decreases towards maturity phase.

B. Factors affecting water application losses

There are huge water losses in conveyance and distribution. It is due to poor land development, unlevelled fields and high seepage loss from unlined uncleaned channels (Table 8.2), deep percolation due to over irrigation, surface run-off inefficient irrigation methods and non-conjunctive use of surface and ground water. While estimating the losses in irrigation it was found that in many cases only one half of the water delivered at the head of the source reached the field and even this was not used properly by the crop.

Table 8.2: Seepage loss in the field channel in summer at Rahuri, Maharashtra

Type of channel	Running length (m)	Average slope (%)	Total wetted surface area (m^2)	Seepage loss ($l/hr/m^2$)
Unlined uncleaned	250	0.55	269	24.78
Unlined cleaned	250	0.40	275	15.78
Lined	250	0.38	153	10.70

Source: Batta,1997

8.2 **Irrigation requirement (IR)**

It refers to the quantity of water required to be applied for successful growth of crop, exclusive of rainfall, ground water contribution and other natural sources. The exact IR can be calculated by considering irrigation efficiency.

IR = NIR + application losses

IR is also called as gross irrigation requirement.

Net irrigation requirement: It is the amount of water required to bring the moisture level in the effective root zone of the soil up to field capacity.

NIR = WR – (ER+GWC+SWB)

Where, WR = Water requirement

ER = Effective rainfall (Normally in irrigated sugarcane areas, rainfall is assumed to be 70% effective)

GWC = Ground water contribution

SWB = Stored water balance

The very climatic pattern of this country makes irrigation indispensable for sugarcane production. Annual rainfall is mostly concentrated in four months from June to September. In other months, there is inadequate or no rainfall. During the dry season irrigation assures the security of cropping while during the wet season irrigation is necessary when dry spells occur.

Sugarcane, being a year long crop, passes through all the three seasons (Summer, Winter and Rainy) in a year. Generally, formative phase of the crop coincides with the summer, grand growth phase with rainy and maturity phase with winter season. During rainy season, depending upon the amount and distribution of rainfall, a part or whole of the water requirement of the crop is fulfilled. During the other two seasons (Summer and Winter) crop depends fully on irrigations. Formative phase of the crop which coincides with summer season is the most critical phase so far as irrigations are concerned. Inadequacy of moisture during this period to any degree causes a major loss in sugarcane yield and quality. Thus, crop is to be irrigated frequently because of the highest evaporative demand due to high temperature and extremely low air humidity. The frequency of irrigations is rather less in winter because of low evaporative demand and well established root system of the fully grown crop during maturity phase.

Extent of irrigational facilities and rainfall pattern in sugarcane growing areas: Out of 142.2 million hectares of net sown area in India, 47.4 million hectares (33.3%) are irrigated by different sources (Table 8.3). Sugarcane is primarily an irrigated crop and 84.4% of its area receives irrigations. However, the extent of irrigated area differs from state to state (Table 8.4). It is grown almost entirely on irrigated lands in the states of Andhra Pradesh, Gujarat, Maharashtra, Karnataka, Tamil Nadu, Haryana and Madhya Pradesh. Punjab, Rajasthan, Uttar Pradesh and Orissa also have fairly large percentage of sugarcane area under irrigation.

There are parts of India where sugarcane is also grown without irrigation in large areas as found in north Bihar, West Bengal, Assam and Kerala. In a study in north Bihar, irrigation in dry hot months did improve yield. This could be a lesson for other states. It is understandable that when grown under unirrigated conditions, there might be occasions when soil moisture status might fall to sub-optimal levels. Such situations could be corrected if irrigation facilities were available. Irrigation would thus safeguard higher yield.

Total annual rainfall and yearly rainfall patterns under different rainfall sub-divisions of sugarcane growing areas of India are given in Table 8.5.

Table 8.3: Per cent area irrigated by different sources in India

Source	Irrigated area (%)
Canals	35.9
Tube wells	30.6
Other wells	20.1
Tanks	6.5
Other sources	6.9

Table 8.4 : Percentage of net irrigated area in the States and irrigated area under sugarcane crop in different States of India (1990-91)

State	Net irrigated area (%)in the state	Irrigated area (%) of sugarcane
	Tropical India	
Andhra Pradesh	39.1	99.0
Gujarat	26.8	100.0
Maharashtra	11.4	100.0
Karnataka	20.4	96.6
Tamil Nadu	42.5	100.0
Kerala	13.6	25.0
Orissa	30.7	79.6
	Subtropical India	
Punjab	92.7	95.0
Haryana	72.7	100.0
Rajasthan	23.8	91.3
Madhya Pradesh	22.1	97.0
Uttar Pradesh	60.9	83.9
Bihar	43.5	14.1
West Bengal	35.8	16.7
All India	33.3	84.4

Source: Anonymous (1994). Water and related statistics, Statistics Directorate, Irrigation Management Organisation, Central Water Commission, pp.65-74.

Table 8.5 : Rainfall pattern in different rainfall sub-divisions of India

Sub-division	Total annual rainfall (mm)	Per cent of annual rainfall			
		June-Sept.	Oct-Dec.	Jan-Feb	March-April
Tropical India					
Coastal Andhra Pradesh	1063	58.2	30.2	2.1	9.5
Telangana	960	81.7	11.0	1.2	6.0
Rayalaseema	717	53.7	33.3	1.8	11.2
Gujarat region	998	95.8	3.1	0.3	0.8
Saurashtra	577	94.5	3.8	0.3	1.4
Madhya Maharashtra	948	83.2	11.5	0.6	4.6
Marathwada	841	84.5	10.2	0.8	4.4
Vidarbha	1133	87.3	6.6	1.9	4.1
Coastal Karnataka	3482	87.2	7.6	0.1	5.0
North interior Karnataka	736	70.6	17.5	0.7	11.0
South interior Karnataka	1156	66.2	19.4	0.7	13.8
Tamil Nadu	1008	32.6	47.9	4.9	14.6
Kerala	3070	69.4	16.6	1.0	13.1
Orissa	1489	79.5	10.3	2.5	7.7

Sub-division	Total annual rainfall (mm)	Per cent of annual rainfall			
		June-Sep.	Oct-Dec.	Jan-Feb	March-April
Subtropical India					
Punjab	643	77.4	6.2	8.4	7.9
Haryana	621	83.1	4.8	6.4	5.6
West Rajasthan	356	88.2	3.9	2.5	5.3
East Rajasthan	772	92.5	3.5	1.7	2.3
West Madhya Pradesh	1059	90.3	5.7	2.1	2.0
East Madhya Pradesh	1344	87.9	5.6	3.0	3.4
East Uttar Pradesh	1022	87.9	5.7	3.4	3.0
West Uttar Pradesh	904	86.0	4.8	4.8	4.5
Bihar Plateau	1331	80.8	8.3	3.1	7.8
Bihar Plain	1194	85.2	6.1	2.8	5.9
West Bengal	1464	76.3	9.9	2.5	11.3
Assam	2836	64.8	7.5	1.9	25.8

Source: Anonymous (1992). Indian Agriculture in brief 24th Edition, 1992, Directorate of Economics and Statistics, Department of Agriculture & Cooperation, Ministry of Agriculture, Govt. of India.

8.3 Irrigation scheduling

In order to get better growth of sugarcane crop irrigation has to be given at a time when soil moisture is not sufficient to cope up freely with the plant water demands. There are different approaches for deciding when to irrigate. These include:

i) Visual symptoms of crop

ii) Depth-interval yield approach

iii) Soil moisture deficit approach

iv) Sheath moisture approach

v) Climatological approach

i) Visual symptoms of crop: Using plant appearance as a criterion for scheduling irrigation is an old age practice being used by the Indian farmers. The success of this practice depends on the nature and accuracy of the observations. Morphological indices such as alterations in the colour of the leaves, noticeable drooping of leaves, loss in turgidity may be used in determining watering times in sugarcane. Under water deficit conditions, the fresh green colour of the leaves changes to a more bluish and ultimately they turn up to dark bluish yellow and subsequently dry up. The first symptom of change of colour from fresh green to bluish tint may be considered as an index for watering the cane. However, delaying irrigation until the plant appearance points to water deficiency in the soil is a harmful practice. By the time these phenomena occur, the plants suffer irremediable damage and cannot yield a maximum harvest. It has also to be kept in mind that sharp changes in external environment such as rise in air temperature or a strong wind may make leaves wilt even in the presence of sufficient soil moisture. Nevertheless, the simplicity of the method has led to its extensive use in irrigation practice for estimating the watering times of sugarcane crop. The observer must be on the alert for such symptoms if he is to correct water deficiency in its early stages without waiting for wilting or alteration in colour, which are always a late sign of advanced metabolic dysfunction.

ii) Depth-interval yield approach: In earlier studies, irrigation scheduling was suggested in terms of days' interval between irrigations (Table 8.6 & 8.7). In this approach, random depth and interval of irrigations were fixed on certain approximations and experiences. The purpose has been to fix interval and depth of irrigation at a longest possible interval before the crop suffers a permanent set-back through moisture stress. Such recommendations are not applicable under wide range of soil and climatic situations as time intervals are influenced by soil type, moisture retention capacity and climatic factor. Thus it leads to improper time, depths and intervals exposing crop to either excess or deficit of water.

Table 8.6: Irrigation numbers* and interval between two irrigations (days) as ascertained on the basis of depth-interval approach for different States of subtropical India

Particular	States							
	Punjab	Haryana	Rajas-than	U.P.	Bihar	West Bengal	Assam	M.P.
Irrigation numbers*								
Formative phase	8-10	8-10	10-12	4-6	4-6	3-4	3-4	10-12
Maturity phase	2-4	2-4	3-5	1-2	1-2	2	2	3-5
Total	11-15	11-15	14-18	6-9	6-9	6-7	6-7	14-18
Irrigation interval								
Formative phase	10-15	10-12	7-15	15-20	15-20	15	15	10
Maturity phase	25-30	25-30	20-25	30	30	30	30	15

*Also includes one preplanting preparatory irrigation for plant cane/ an irrigation given at initiation stage in ratoon cane to stimulate bud sprouting.

(Note: In case of long gap of rains during the *monsoon*, the crop should be irrigated as per need. Further, in the years of increased sugarcane production the mills are forced to extend their crushing period to hotter months of April, May and sometimes even in June. A steep fall in sugar recovery and cane yield after March, due to over maturity and desiccating winds, is a common observation. Cultivators generally do not irrigate the mature crop which further aggravates the situation. Studies at Jalandhar (Kapur and Kanwar, 1984) indicated that under such situations, the application of irrigation water at fortnightly interval to the standing crop helps in reducing the fall in juice quality and yield.by 50% in April and May. However, during June the effect of irrigation was low).

iii) Soil moisture deficit approach: Moisture is not available readily and equally between the range of field capacity (FC) and permanent wilting point (PWP) has given the concept of soil moisture depletion approach. The available water in soil is generally considered as that held between the FC and PWP. But this water is not available to the plants at the same ease throughout the available range because of variation in moisture retention force. Beyond a particular range, the plant has to put forth more osmotic pressure to draw water from soil which hinders normal growth of the sugarcane. The plant growth is adversely affected after a certain level of moisture depletion and that level occurs much before the PWP. The safe limits of allowable water depletion for different growth stages of sugarcane have been determined through field experimentation all over the country. In most of the experiments, irrigation at 50% depletion of available soil moisture in the root zone at 15-30 cm soil depth during formative phase (first four months of the crop) and at 75% depletion during maturity phase (last 3 months of a 12-month crop) lead to a maximum cane yields.

Table 8.7: Interval between two irrigations and total number of irrigations for sugarcane as ascertained on the basis of depth-interval approach for tropical India

State	Irrigation interval (days)	Total number of irrigations	Reference
Maharashtra		30-35	Hapse, 1991
Summer	8-10		
Winter	14-15		
Rainy season	18-21		
Tamil Nadu		30-35	Kailasam, 1995
Germination (upto 35 days)	7		
Tilling (36-100 days)	10		
Grand growth (101-270 days)	7		
Maturity	15		
Karnataka		30-35	Khandagave & Kagi,1998
February-March	10		
April-May	8		
May onwards	12		
Andhra Pradesh		25-30	Mathur & Behari, 1980
Formative phase	6-7		
Maturity phase	21		
Gujarat	7-10	20-30	
Orissa	7-12	-	
Kerala	-	8-10	

The deficit is calculated by the formula:

$$d = \frac{FC - A}{100} \times BD \times D$$

where, d = Moisture deficit in a soil layer (cm)

FC = Field capacity (%)

A = Actual moisture content in the soil (%)

BD = Bulk density of the soil

D = Depth of soil layer (cm)

The net deficit in the root zone thus worked out is corrected by applying required depth of irrigation water.

For quick measurements of soil moisture, instruments like tensiometers, gypsum blocks and neutron probe may be used. The degree of depletion is measured with differential criteria such as percentage availability, tensions and resistance etc.

The technique (approach), however, is satisfactory for practical purposes. But it needs costly equipments such as tensiometers and resistance units for soil moisture control. Soil sampling to root zone depth is also laborious and time consuming. This involves measurement of soil moisture at a number of time throughout the crop growth period. The practical utility of irrigating the crop at specified soil moisture regime has a remote possibility under Indian conditions. The reasons are many. Firstly, water is not at command of the farmers due to closure of canals. Secondly, farmers are not well trained to handle sophisticated equipments.

However, with all the limitations, tensiometer can play a vital role in scheduling irrigation to sugarcane grown on sandy or sandy loam soils and irrigated through wells and reservoirs. In coarse textured soils most of the water is held at low tensions and more than 75% available soil water is utilized at a tension of 0.85 atmosphere. Because of narrow working range (up to 0.8 bar) tensiometers cannot be used for situations where irrigations are delayed up to tensions greater than one bar. Ray *et al.* (1964) and Prasad and Parashar (1986) reported that maintaining a soil moisture regime of 0.7 atmosphere tension measured at 22.5 cm soil depth throughout the crop life-cycle was adequate for growing a successful crop of sugarcane under Delhi conditions in sandy loam soil. The crop required 6-8 irrigations before the *monsoon* period and 2-3 irrigations during the post-*monsoon* period.

iv) Sheath moisture approach: The importance of plant-water status in relation to growth had led to measurement of tissue moisture. It has been found that sheath moisture level may be used as an indicator for detecting water deficit in sugarcane plant for scheduling irrigation. In this method, the moisture per cent in the leaf sheaths (3-6 leaf sheaths from the top) of sugarcane plant is determined periodically and fall in moisture percentage below critical level is a "call" for irrigation. Studies conducted by Lakshmikantham *et al.* (1962) at Anakapalle suggested that soil moisture regime should be so adjusted that the moisture content of the 3-6 leaf sheath does not drop below 85% during formative phase of the crop growth (first four months). This should be lowered gradually afterwards by withholding irrigations and should reach to 73% at the time of harvesting. However, it will also vary up to a certain extent with cane varieties. Any attempt to ripen the crop prematurely, by bringing about a drastic fall in the sheath moisture, has been observed to be definitely harmful, resulting in poor yields. It is, therefore, advised that irrigation control measures can be resorted to with advantage, after the age of 10 months of planting the cane.

v) Climatological approach: Evapotranspiration (ET) from the cropped area has a very close relationship with the evaporation of water from open pan evaporimeter. Thus, the use of open pan evaporimeter has become an important tool for scheduling irrigation in field crops. The rate of evaporation from a free water surface depends upon energy input, duration of energy input and the factors associated with the removal of evaporated water vapour. Evaporation from open pan, therefore, integrates the effects of temperature, radiation, wind current and humidity and is very easy to measure. Scheduling irrigation using open pan is simple, rapid, and reliable and has high extension value. ET values can also be computed from various empirical formulae. However, pan evaporimeter is considered to be more accurate in estimating short term fluctuations in ET than the empirical formulae that depend on fewer of the climatic factors involved. The Standard U.S. Weather Bureau Class-A pan evaporimeter is widely used to measure evaporation under field conditions for scheduling irrigation to sugarcane crop. In this method, after a specified amount of water evaporation from pan evaporimeter, a measured quantity of water is applied to the field, sufficient to fill the soil reservoir making allowance for reasonable efficiency. Ratio of crop ET to pan evaporation is low in early stage of crop growth period, higher in grand growth phase and again low in senescent phase.

A lot of research work has been carried out on scheduling of irrigation based on open pan evaporation values and at most of the places in India, irrigations applied at IW/ CPE (irrigation water / cumulative pan evaporation) ratio of 0.8–0.9 have resulted in significant increase in cane yield as compared to a IW/CPE ratio of 0.5 during pre-*mosoon* period (Tables 8.8 to 8.10)). However, under *tarai* conditions of

Table 8.8: Cane yield as influenced by irrigations applied at different levels of water evaporation from open pan and IW/CPE ratios at different places in India (Depth of water 80 mm/irrigation)

Irrigation at		Cane yield (t/ha)	
Open pan evaporation (mm)	IW/CPE ratio	1982-83	1983-84
Karnal(Haryana)			
160	0.5	65.3	71.1
114	0.7	73.0	78.1
88	0.9	85.4	85.1
CD(0.05)	-	12.2	6.3
Pantnagar (Uttaranchal)			
160	0.5	81.9	48.7
114	0.7	84.7	46.8
88	0.9	81.2	48.3
CD(0.05)	-	NS	NS
Jalandhar (Punjab)			
160	0.5	67.1	46.4
114	0.7	71.0	48.6
88	0.9	74.9	52.3
CD(0.05)	-	2.7	3.4

Source: Ann. Rep. AICRPS (1982-83) p.47, (1983-84) pp. 33-42

Table 8.9: Cane yield as influenced by irrigation applied at different IW/CPE ratios in various States of India.

Irrigation at IW/CPE ratio of	Cane yield (t/ha)		Soil type	Reference
	1988-89	1989-90		
Karnal (Haryana)				
0.5	53.6	52.0	Sandy	Kumar &
0.8	65.4	58.3	loam	Srivastava,1991
CD(0.05)	10.2	6.0		
Muzaffarnagar (Uttar Pradesh)				
0.5	67.3	49.2	Sandy	Malik
0.8	71.2	51.0	loam	*et al.*,
CD (0.05)	0.8	NS		1996
Sehore (Madhya Pradesh)				
0.5	74.4	68.3	Clay	Sharma
0.8	83.3	75.5	loam	& Verma
CD (0.05)	4.7	4.7		1996
Lucknow (Uttar Pradesh)				
0.5	63.4	51.9	Loam	Ann. Rep. AICRPS
0.8	68.3	55.5		1988-89,pp 20-25
CD (0.05)	2.5	2.4		1989-90, pp 35-42
Pantnagar (Uttarakhand)				
0.5	93.6	-	Clay	-do-
0.8	88.6	-	loam	
CD (0.05)	NS	-		
Padegaon (Maharashtra)				
0.5	121.6	83.8	Clay	-do-
0.8	134.6	88.1		
CD (0.05)	5.3	1.3		

Irrigation at IW/CPE ratio of	Cane yield (t/ha) 1988-89	1989-90	Soil type	Reference
		Navsari (Gujarat)		
0.5	72.2	92.7		-do-
0.8	97.5	103.0	Clay	
CD (0.05)	4.8	2.9		
		Coimbatore (Tamil Nadu)		
0.5	56.0	-		-do-
0.8	71.8	-		
CD (0.05)	2.9	-		
		Cuddalore (Tamil Nadu)		
0.9	105.8	122.6	Clay	Pandian
1.1	109.5	133.0	loam	*et al.*,
CD (0.05)	NS	NS		1992

Table 8.10: Cane yield as influenced by irrigations applied at different IW/CPE ratios in sandy loam soil at Jalandhar, Punjab (Depth of water 75 mm/irrigation)

Irrigation at pan evaporation (mm)	IW/CPE ratio	Cane yield (t/ha) 1992-93 1993-94	1993-94	Mean
200	0.375	58.8	43.4	51.1
100	0.75	67.3	57.4	62.4
50	1.50	69.4	61.2	65.3
CD (0.05)	-	4.8	11.1	5.8

Source: Toor *et al.*,1999

Pantnagar where shallow water table conditions exist, irrigating sugarcane crop at IW/CPE ratio of 0.5 was adequate to meet the crop water need.

Subramanian *et al.* (1989) and Thanki *et al.* (2000) observed that irrigating sugarcane crop with 0.75, 0.75, and 5.0 IW/CPE ratio at tillering, grand growth and maturity phases, respectively, gave the highest yield of cane. It indicates that moisture stress during juvenile and grand growth phases of sugarcane crop reduces the yield significantly. But at maturity stage the crop can withstand moisture stress to certain extent (Table 8.11).

Table 8.11: Cane yield as influenced by irrigations applied at various IW/CPE ratios at different growth stages of sugarcane in clay soil at Navsari, Gujarat (Mean of data from 1989-90, 1991-92; 1992-93) (Depth of water 80 mm/irrigation)

IW/CPE ratio at different crop growth stages			Cane yield (t/ha)
Juvenile	Grand growth	Maturity	
0.75	0.75	0.75	107.2
0.50	0.50	0.50	86.7
0.75	0.75	0.50	104.6
0.50	0.50	0.75	89.2
0.75	0.50	0.75	91.9
0.50	0.75	0.50	96.0
CD(0.o5)			8.7

Source: Thanki *et al.*, 2000

8.4 **Depth of irrigation (Amount of water per irrigation)**

If the soil is regarded as a water reservoir to meet water demands of crop, the volume of water it can hold up to its effective root depth is very important for the quantity of water to be applied per irrigation. It depends upon the effective root depth (Fig. 8.1) and soil texture. Sugarcane roots are mostly distributed in the upper 60 cm soil layer. It has been observed in various studies that vast majority of fibrous roots,

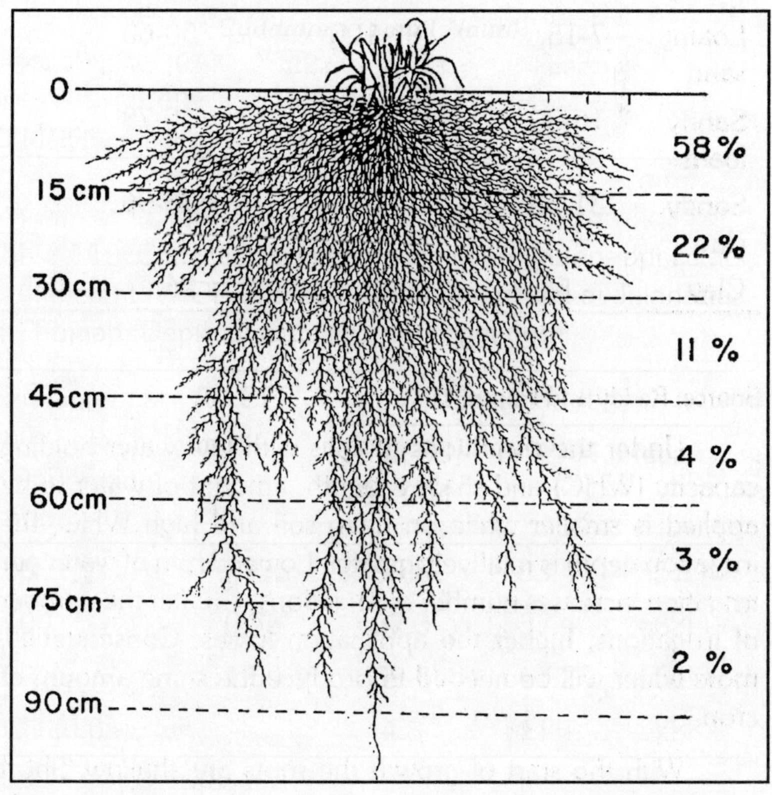

Fig. 8.1: Pattern of root distribution of sugarcane

which are most active in water absorption occur in the upper 30 cm soil. Available moisture content values for different textural classes are given in Table 8.12a.

Table 8.12a: Soil texture and depth in relation to available moisture content (AMC)

Textural class	Clay content (%)	Expected AMC	
		mm/m	mm/60 cm
Sandy	0-6	80	50
Loamy sand	7-15	81-100	50-60
Sandy loam	16-30	101-130	60-78
Sandy clay	31-55	131-160	78-96
Clay	> 55	160	96

Source: Reddy and Hapse, 1986

Under the conditions of soils with low water holding capacity (WHC) and shallow depth, amount of water to be applied is smaller while on deep soil and high WHC, the irrigation depth is relatively greater. Lower depth of water per irrigation increases number of irrigations. Greater the number of irrigations, higher the application losses. Consequently, more water will be needed to produce the same amount of crop.

With the start of growth the roots are shallow. Short irrigational intervals with less amount of water are suitable for the early periods of growth. Later, the amount of irrigation water must be increased to match the high WR.

The water supply in deficit or in excess is not desirable. Deficit supply leads to underirrigation which results in water stress and loss of yield while excess application impedes aeration and removes plant nutrients from active root zone by deep percolation. In the long run, problems of salinity and waterlogging may develop due to rise in ground water table.

It has been a common experience that on account of ample availability of canal water and ignorance about the exact amount of water to be applied, the cane growers are often tempted to use excessive amount of water which not only lead to considerable wastage of water, but in course of time results in serious damage to soil from waterlogging and salt efflorescence. Moreover, night irrigations are predominant for sugarcane crop. Night irrigations are always associated with excess watering. Excessive use of water should always be discouraged.

The amount of water to be applied (deficit) may also be calculated by the formula:

$$d = \frac{FC - A}{100} \text{ X BD X D}$$

Where, d = moisture deficit in a soil layer (cm)

FC = Field capacity (%)

A = Actual moisture content in the soil (%)

BD = Bulk density of the soil

D = Depth of soil layer (cm)

The net deficit in the root zone thus worked out is corrected by applying required amount of water considering irrigation efficiencies in view.

The knowledge of water distribution pattern after irrigation in various soils is also important as it gives an idea of controlling the speed of irrigation water. Soils with low rate of water intake (infiltration) must be irrigated slowly to ensure adequate penetration. Rate of water intake by a soil varies as per soil texture (Table 8.12b).

Table 8.12b: Infiltration rate as influenced by soil textural class

Soil type	Infitration rate (mm/hr)
Sands	20
Sandy and Silty	10-20
Loams	5-10
Clay	1-5
Sodic clay	1

8.5 **Depth of water table**

Information available in the country on the effect of water table depth on sugarcane crop is scanty. Depth of water table may affect the sugarcane crop in two ways:

i) Adversely affecting growth and yield

ii) Contributing to total evapotranspiration (ET)

i) Adverse effect on growth and yield: Very shallow depth (< 60 cm) of water table causes greater reduction in growth and yield of sugarcane (Table 8.13)

In studies of Gosnell (1971) cane was grown for three years in deep containers in which the water table was maintained at constant depths of 25, 50, 75, 100 and 125 cm. The 25-cm water table inhibited germination at planting and ratooning and caused large reduction in stalk number, height, cane yield, root weight and sucrose per cent. The 50-

Table 8.13: Yield of sugarcane grown under different depths of water table at Lucknow

Ground water table depth (cm)	Cane yield (t/ha)
20	78.5
40	94.2
60	98.8
CD(0.05)	7.4

Source: Gupta and Yadav, 1993

cm water table gave intermediate results, and no differences in growth were observed between 75, 100 and 125 cm water tables, which gave the best results. These effects become progressively more severe from plant crop to second ratoon. Salinization of the soil occurred about 20-30 cm above the water table in all the treatments, it was not serious in the 100- and 125-cm water table treatments, but fairly high values of electrical conductivity and sodium absorption ratio of the soil saturation extract were observed in the 25-, 50- and 75-cm water table treatments.

ii) Contribution to total ET: Shallow depth (60-100 cm) of water table could be extremely useful in providing part of the water needed by the sugarcane crop and the interval between two irrigations could be lengthened. Water table at a depth of about 1 meter contributed 65.13% towards total ET of sugarcane grown on red sandy loam soils of Mandya (Karnataka) and had no deleterious effect on growth, yield and quality (Hunsigi and Srivastava, 1977).

It may be concluded that the very shallow water table depth (< 60 cm) causes adverse effect on cane growth and yield while a water table depth of 60-100 cm can contribute to total ET. Water table beyond one meter neither contributes to ET nor affects the sugarcane growth and yield adversely.

8.6 **Methods of irrigation**

The irrigation methods adopted in sugarcane crop may be broadly grouped into three categories:

8.6.1 Surface

8.6.2 Overhead or Sprinkler

8.6.3 Drip or trickle irrigation

8.6.1 **Surface irrigation**: Surface irrigation is the most common method of applying water to sugarcane in India. In this method, water is brought to the fields by a series of open unlined channels and finally diverted either to the entire field, check basins, border strips or furrows. The application of water by surface irrigation requires smooth land surface so that the farmer can control the water as it flows across the field. In order to put the land surface to a smooth uniform slope, precision land grading is necessary. The application of water through this method is comparatively easier and economical but there are certain disadvantages associated with it. These are:

- Low irrigation efficiency ranging between 30-40% due to greater loss of water through evaporation, surface run-off, deep percolation and seepage.
- Uneven distribution of water in fields due to uneven gradient which leads to under-irrigation of upland spots or overirrigation in low lying spots resulting in poor cane yields.
- Decrease in effective area of land by 10-15 % which is used for irrigation layouts.

- Undesirable weed growth.
- Exposure of crop to cyclic changes of water and air stresses. The crop has to face the air stress condition immediately after irrigation and moisture stress before the irrigation. Air and water stress conditions not only affect the plant growth, root respiration but also adversely affect the microbial activities which are doing excellent job for nutrient transformation and availability.
- Loss of nutrients due to deep percolation and seepage.

In surface irrigation, different layouts are made in the field in order to improve the irrigation efficiency. Some of the layouts of surface irrigation are:

i) Check basins: In this method water is applied to small and relatively levelled plots surrounded by ridges. Water is conveyed to the fields by supply ditches (channels) and field ditches. There is generally one field ditch for two rows of check basins. The checks are generally kept rectangular for efficient irrigation. The ridges should be made along the contours of the field. The size of the checks is based upon the gradient of the field. However, this system does not serve the purpose in areas where sugarcane is planted in ridge and furrow system.

ii) Border strips: In this method the field is divided into a series of long strips called borders that are separated by low ridges. The borders have gentle downwards uniform gradients and are level cross-wise. Normally, the direction of the strip remains in the direction of the greatest slope. Water is applied on to the upper end of each strip and slowly flows towards the lower end in a thin sheet. The infiltration opportunity criterion for the uniformity of application and adequacy of irrigation is demarcated by the time when water advances to and recedes from a point. Thus the time-distance relationship of water front advance and tail water recession are major factors for determining border irrigation efficiency.

iii) Furrow irrigation: In this method water is applied by running small streams in furrows made in between the cane rows and allowed to soak down in the root zone of the plants. The direction of furrows should be across the gradient of the field. The furrow length is determined by soil permeability and steepness of the slope. On slopes less than one per cent the furrow runs should be longer and on steeper slopes the runs must be shorter. Longer lengths of run are feasible on heavy clay soils. The furrows may be opened with a heavy ridger.

This method is more efficient as compared to other surface irrigation methods as water in furrows comes in direct contact with only $^1/_3$ to ½ land surface thereby reducing evaporation losses. Puddling and soil crusting are also observed to be less.

To have water economy or in situations of water scarcity farmers can switch over to alternate or skip-furrow irrigation instead of every-furrow irrigation. As is indicated by the term skip or alternate, in this method of irrigation, instead of irrigating all the rows and inter-row spaces, one row is skipped (avoided or left out) for irrigation purposes, and irrigation is given to alternate rows (Fig. 8.2). Sometimes, alternate-furrow irrigation results in greater crop yields Table 8.14) because of a better soil-water relationship. Alternating-alternate-furrow irrigation may also be practiced. In this method, water is applied in alternate furrows in successive irrigations. At the every next irrigation, water is applied in furrows that did not receive irrigation at the preceding irrigation. No differences

Fig. 8.2 : Lay-out for skip-furrow method of irrigation in sugarcane field

were observed between alternate and alternating alternate-furrow irrigation. Skip/alternate furrow irrigation saves 30-40% water without reduction in cane yield (Tables 8.15 to 8.17).

Table 8.14: Cane yield under alternate-furrow irrigation in black soil at Belgaon, Karnataka

Irrigation method	Cane yield (t/ha)		Mean
	1968-69	1969-70	
Every furrow	112.5	83.3	97.4
Alternate-furrow	126.9	83.3	105.0
Mean	119.7	83.3	-

Source: Hunsigi and Parmeshwarappa, 1975

Table 8.15: Cane yield, water saving and WUE under skip-furrow irrigation in alluvial entisol at Lucknow, Uttar Pradesh

Irrigation method	Cane Yield (t/ha)	Water used (ha-cm)	Water saving (%)	WUE (t/ha-cm)
Every furrow	63.2	25.0	16.7	2.53
Skip-furrow	77.1	18.8	37.3	4.10
Check-basin	74.3	30.0	-	2.48
CD(0.05)	9.0	-	-	-

Source: Prasad *et al.*, 1980

Table 8.16: Cane yield, water saving and WUE in plant and ratoon crops of sugarcane under skip-furrow irrigation in alluvial entisol at Lucknow, Uttar Pradesh

Irrigation method	Cane yield (t/ha)	Water used (ha-cm)	Water saving (%)	WUE (t/ha-cm)
Plant crop (mean of 1978-79 and 1979-80)				
Skip-furrow	66.9	22.5	25.0	2.97
Check basin	66.3	30.0	-	2.31
CD(0.05)	NS	-	-	-
Ratoon (1980-81)				
Skip-furrow	75.9	30.0	33.3	2.53
Check basin	71.2	45.0	-	1.58
CD(0.05)	NS	-	-	-

Source: Prasad *et al.*, 1987

Table 8.17: Cane yield, water saving and WUE under alternate-furrow irrigation at Seorahi, Uttar Pradesh (Mean of data from 1989-90 and 1990-91)

Irrigation method	Cane yield (t/ha)	Water* used (ha-cm)	Water saving (%)	WUE (t/ha-cm)
Every furrow	97.3	142.7	24.4	0.682
Alternate-furrow	94.0	126.0	33.2	0.746
Check basin	89.3	188.7	-	0.473
CD(0.05)	5.4	-	-	-

Source: Singh *et al.*,1994 *Inclusive of rainfall

8.6.2 Overhead irrigation/Sprinkler irrigation: In this method, water is sprayed over the crop just like rain through sprinkler nozzles (Fig. 8.3). When water is applied under the sprinkler system, the rate of water application is constant, and thus the relation between cumulative intake *vs.* time is a straight line as long as ponding does not occur. Ponding does not occur if the water application rate does not exceed the infiltration rate of the soil.

Merits

- Since water is conveyed through pipes and applied as per the requirements of the crop, the water conveyance and deep percolation losses could be avoided. So irrigation efficiency of this method is higher (80%) than surface irrigation (50%).
- It is a good method for undulating topography and porous soils.

Fig. 8.3 : Sprinkler irrigation in sugarcane

- Desired soil-air-water relationship with excellent soil physical environment could be maintained.
- Cultivable area is saved due to elimination of irrigation ditches.
- Non-interference with traffic movement due to absence of open ditches.

Demerits

The use of sprinkler irrigation in sugarcane agriculture in India is restricted due to the operational difficulties, involvement of high investment, high-energy need and relatively less water saving.

- If proper care in the design of field layout and shifting

of sprinklers is not taken the corners of the field remain unirrigated.

- The spray is invariably affected by the wind speed and direction. If wind speed exceeds 10 km/hr the spray is directed on one side making uneven distribution of irrigation water throughout the field.
- As sugarcane crop grows, the height of risers has to be adjusted accordingly.
- The operation of sprinkler irrigation system requires 2.5 to 3.0 kg/cm^2 pressure and therefore more electrical energy is needed. Failure of electricity is also a problem.
- In summer, due to high temperature, nearly 10-15% of applied water is evaporated before it is received on crop or ground surface.

Experimental findings: The trials conducted at VST, Pune and various cooperative sugar factories in Maharashtra (Hapse, 1991) and at Navsari, Gujarat (Table 8.18) revealed that 15 to 20% increase in cane yield and 30-40% saving in irrigation water could by obtained by sprinkler irrigation in sugarcane.

8.6.3 **Drip irrigation**

Drip/trickle irrigation is an advanced method of supplying required quantity of water to the root zone through a network of tubing and drippers/emitters/nozzles placed along water delivery. It is a novel concept involving precise control and manipulation of soil moisture both temporally and spatially which results in water economy and increased growth, yield and quality of the crop.

In this system, water is supplied from a point source. The supply of water is made through lateral pipes placed along

Table 8.18: Effect of different methods and levels of irrigation on cane yield and WUE in heavy black soil at Navsari, Gujarat (Mean of data from 1988-89 and 1990-91)

Irrigation method	IW/CPE ratio	Number of irrig-ations	Water applied (ha-cm)	Cane yield (t/ha)	Water saving (%)	WUE (t/ha-cm)
Sprinkler	0.30	8	40.0	94.7	45.5	2.367
	0.45	11	55.0	97.2	37.5	1.768
	0.60	14	70.0	94.8	20.5	1.355
	0.75	17	85.0	89.7	3.4	1.055
Furrow	0.75	11	88.0	79.4	-	0.902

Source: Shrivastava *et al.*, 1993

(Note: Irrigation of 50 mm depth was applied by sprinkler when CPE reached 167, 111, 83 and 67 mm in 0.3, 0.45, 0.60 and 0.75 IW/CPE ratios, respectively. In furrow irrigated plots, 50 mm depth of water was given when CPE reached 100 mm)

the crop rows. The soil wetting pattern is controlled by emitter discharge rate and irrigation regime which is in turn determined by frequency and amount of irrigation and crop water use.

Based on the remarkable response of a single orchard tree to an accidental slow dripping leakage from a pipeline in the soil, the farmers of Israel developed the concept of systematic drip irrigation for a variety of crops. The slow release of water to the root zone, almost at the rate at which the plant can absorb the same, results in bringing about spectacular savings in water requirements. Also, if the plant nutrients are dispensed through irrigation water, the results can be maximized. Drip irrigation system, besides economizing enormous quantities of water, go a long way to create a congenial conditions for the root zone (Fig. 8.4) of the plants for enabling them to maintain sustained growth. The soil structure is left undisturbed, and free aeration of the root zone is allowed even while the irrigation goes on. This results in increased yield and quality.

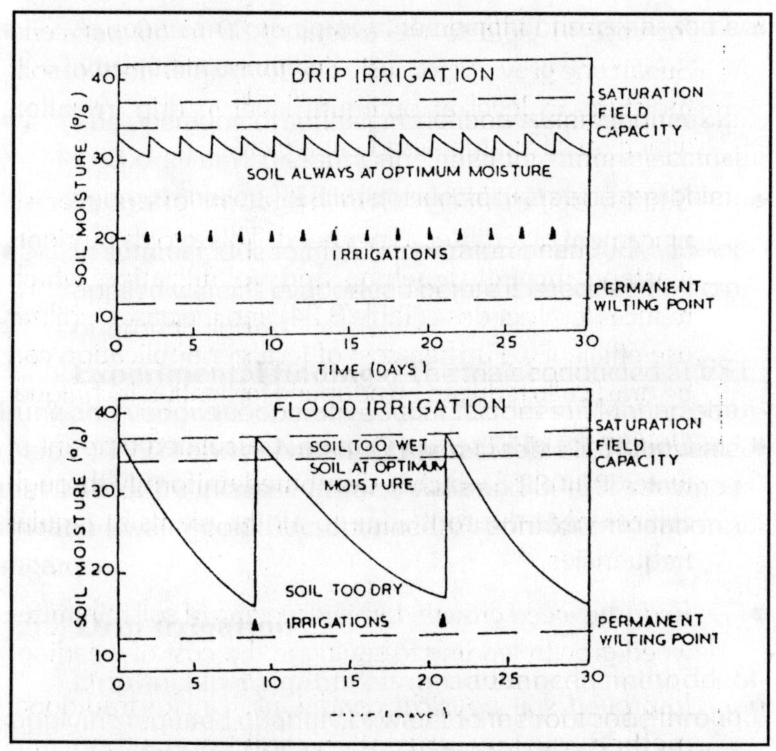

Fig. 8.4 : Typical patterns of soil moisture changes over time for drip and flood irrigation systems

Source : Hapse, 1989.

Advantages of drip irrigation: Drip irrigation system offers several advantages. These are:

- *Water saving from 50 to 75 per cent.* Water is delivered to the root zone through pipes, thus greatly reducing water loss through evaporation, surface run-off and deep percolation (Tables 8.19 to 8.23).
- *Increase in yield to the extent of 30 to 50 per cent.* Sugarcane grows vigorously due to maintenance of soil-moisture-air level at optimum level in drip irrigation which results in increased cane yield.
- *Saving of fertilizers.* Drip irrigation offers precise placement of fertilizer as per need. This avoids nutrients wastage through leaching and volatilization which reduces fertilizer dose (Table 8.24) and increases fertilizer use efficiency. Further, cost of fertilizer application can be drastically reduced by adopting fertigation technique.
- *Uniformity of water application.* Calculated amount of water is applied which is distributed uniformly through orifices near the root zone to all the plants at regular frequencies.
- *Reduced weed growth.* Limited wetting of soil minimizes weed growth leading to saving in the cost of weeding.
- *Improved soil physical conditions.* Unlike traditional method, surface crustation, soil compaction and destruction of soil structure is avoided.
- *Intercultivation may be avoided.* It is not necessary to manipulate the soil physically on and off since optimum soil-moisture -air relationship is automatically maintained if crop is irrigated by drip method. Further, if there are any weeds, they may be controlled by using

Table 8.19: Cane yield, water saving and WUE of sugarcane under drip and furrow methods of irrigation (Mean of data from four locations viz., Kolhapur, Ternanagar, Dhamangaon and Kopargaon of Maharashtra)

Details of treatment		Cane yield (t/ha)	Cane yield increase (%)	Water* used (ha-cm)	Water saving (%)	WUE (t/ha-cm)
Irrigation method	Irrigated at ETc					
Surface drip	1.75	90.2	22.2	151.8	32.3	0.594
	1.50	92.8	25.7	142.4	36.5	0.651
	1.25	93.9	27.2	131.6	41.3	0.713
	1.00	94.3	27.8	121.6	45.7	0.775
Subsurface drip	1.75	90.5	22.6	142.9	36.3	0.633
	1.50	92.9	25.9	134.6	40.0	0.689
	1.25	96.5	30.8	127.6	43.1	0.756
	1.00	90.6	22.8	114.4	49.0	0.792
Furrow Conventional		73.8	-	224.2	-	0.329
CD	(0.05)	6.8	-	-	-	-

Source: Hapse et al.,1991 ETc = Crop Evapotranspiration
*Including effective rainfall

Table 8.20: Cane yield, water saving and WUE under drip and furrow methods of irrigation in clay soil at Padegaon, Maharashtra (Mean of data from 1989-90 and 1990-91)

Irrigation method	Cane yield (t/ha)	Yield Increase (%)	Water applied (ha-cm)	Water saving (%)	WUE (t/ha -cm)
Surface drip	116.5	17.8	176.2	21.4	0.661
Subsurface drip (Biwall)	112.7	14.0	142.6	36.4	0.790
Furrow	98.9	-	224.1	-	0.441

Source: Gholve et al.,1994

Table 8.21: Cane yield, water saving and WUE under different methods of irrigation at Coimbatore, Tamil Nadu (Mean of data from 1977-78 and 1978-79)

Irrigation method	Cane yield (t/ha)	Cane yield increase (%)	Water Applied (ha-cm)	Water saving (%)	WUE (t/ha-cm)
Drip	110.7	+30.4	79.4	53.1	1.394
Skip-furrow	77.6	–8.6	86.9	31.3	0.893
Every furrow	84.9	-	126.5	-	0.671

Source: Sivanappan, 1983

Table 8.22: Cane yield, water saving and WUE under furrow and drip methods of irrigation in heavy black soil at Navsari, Gujarat (Mean of data from 1989-90 and 1990-1991)

Details of treatment		Cane	Yield	Water	Water	WUE
Irrigation method	Irrigation at IW/CPE ratio of	yield (t/ha)	increase (%)	applied (ha-cm)	saving (%)	(t/ha)
Drip	0.3	106.4	34.0	44.0	50.0	2.418
	0.4	117.3	47.7	55.4	37.0	2.117
	0.5	115.9	46.0	62.7	28.8	1.848
	0.6	120.5	51.8	72.0	18.2	1.674
	0.7	108.6	36.8	81.3	7.6	1.336
Furrow	0.75	79.4	-	88.0	-	0.902

Source: Parikh *et al.*,1992

Table 8.23: Cane yield, water saving and WUE in plant and ratoon crops of sugarcane under drip and furrow methods of irrigation in Taiwan.

Crop stand	Irrigation method	Cane yield (t/ha)	Yield increase (%)	Water applied (ha-cm)	Water saving (%)	WUE (t/ha -cm)
Plant crop	Drip	107	21.6	244	40.6	0.438
	Furrow	88	-	410	-	0.214
Ratoon	Drip	72	53.2	244	40.6	0.295
	Furrow	47	-	411	-	0.114

Source: Anonymous, 1978

Table 8.24: Cane yield as affected by different levels of nitrogen applied through drip irrigation system at Pune and Kopargaon, Maharashtra

Treatment	Cane yield (t/ha)		Mean
	Pune	Kopargaon	
125% N through drip	72.7	63.0	67.9
100% N through drip	63.0	74.3	68.7
75% N through drip	68.7	80.2	74.5
50% N through drip	52.4	72.2	62.1
100% N applied conventionally	67:8	70.8	69.3

Source: Hapse, 1991

systemic selective herbicides along with water through drip.

- *Saving on energy requirement.* Energy consumption for pumping of water is reduced from that which is required for sprinkler irrigation. Pumps perform more efficiently at low heads (operating pressure is less than 1.0 atm with drip irrigation compared to 3.0 to 7.5 atm with sprinkler irrigation).
- *Cost on levelling of the land can be minimized.*

Disadvantages: Generally, the drip irrigation method is criticised for high initial investment and maintenance problems.

Overall cost of the system seems high. On small area, drip system is costly (Rs. 50,000 - 55,000 /ha) which is prohibitive for the individual farmer. It is necessary to think about group or cluster farming where the costly items like permanent structure, silting and filter tanks of the system can be shared by the individuals in the cluster.

- *The lateral pipes either crack or the emitters/drippers clog very badly.* It has been observed that subsurface drip system chokes to the tune of 15 to 20% during the first year of installation itself. It has also been observed that while removing and relaying the system in the field for its reutilization, the damage is nearly 20 to 30%. To overcome these problems and to increase the life of the system, flushing arrangements may be added to the system to remove any silt/foreign material in the pipes. Care should also be taken by the farmers in the operation and maintenance of the system. As far as possible, all the pipes, including laterals, may be placed underground.

- *More cane lodging is experienced in drip irrigated crop.* In drip irrigation system, wetting zone is of limited volume and is restricted upto 20-30 cm depending upon the intensity of dripping. This brings restriction on the soil root volume which reduces plant anchorage. To tackle this problem cane tying and propping operation is to be introduced in sugarcane cultivation.

Classification of drip system: Drip irrigation for sugarcane is either done by placing the LDPE/LLDPE (low density polyethylene) lateral tubes with drippers/emitters on the surface (Fig.8.5) between two rows of sugarcane or by burying the biiwall (Fig. 8.6) in the soil. The drip irrigation systems commonly used for irrigating sugarcane are classified as:

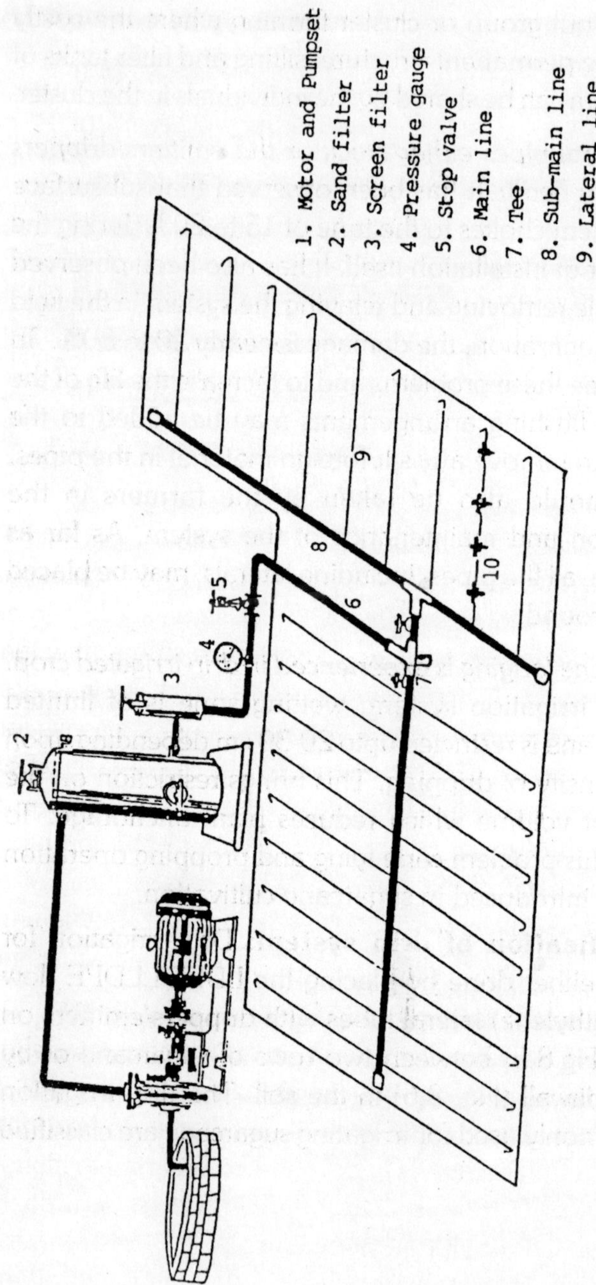

Fig. 8.5 : Schematic sketch - showing surface drip irrigation system

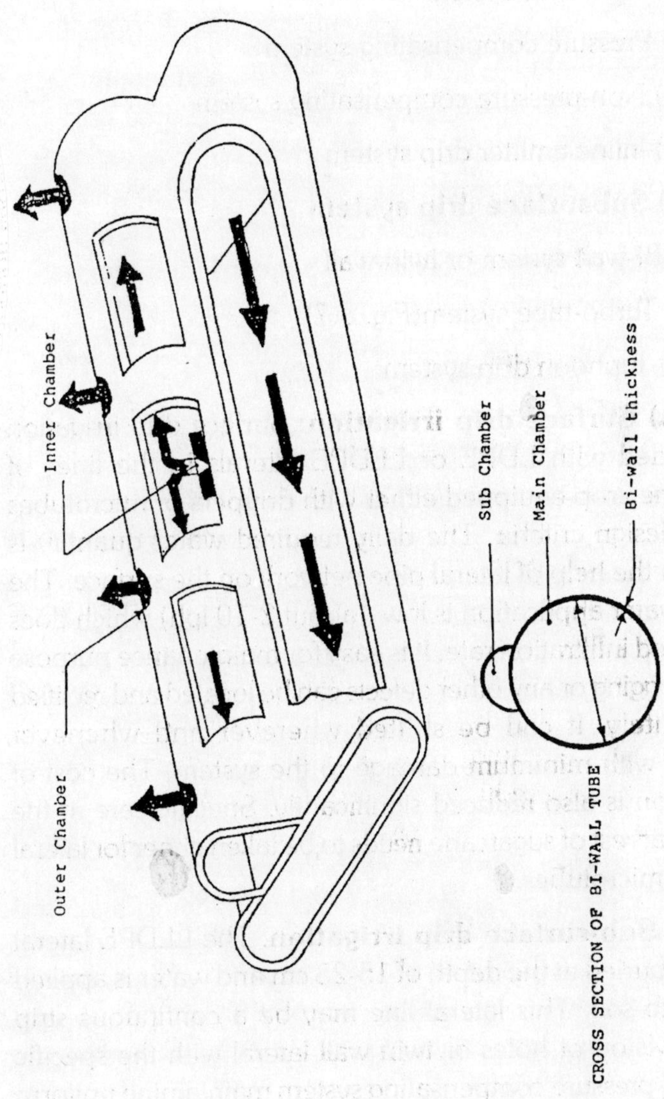

Fig. 8.6 : Bi-wall tube.

A) Surface drip system

i) Micro-tube system

ii) Pressure compensating system

iii) Non-pressure compensating system

iv) Inline emitter drip system

B) **Subsurface drip system**

i) Bi-wall system or twin-wall

ii) Turbo-tape system(Fig. 8.7)

iii) Typhoon drip system

(A) Surface drip irrigation: Surface drip irrigation is provided with LDPE or LLDPE laterals in the lines of sugarcane crop equipped either with drippers or microtubes as per design criteria. The daily required water quantity is met with the help of lateral pipe network on the surface. The rate of water application is low (about 2-10 lph) which does not exceed infiltration rate. It is easy for maintenance purpose as the clogging or any other defects can be located and rectified immediately. It can be shifted wherever and whenever required with minimum damage to the system. The cost of installation is also reduced significantly. Specific care at the time of harvest of sugarcane needs to be taken either for lateral pipes or microtubes.

B) Sub-surface drip irrigation: The LLDPE lateral lines are buried at the depth of 15-25 cm and water is applied in the sub-soil. This lateral line may be a continuous strip with provision of holes or twin wall lateral with the specific benefit of pressure compensating system maintaining uniform discharge throughout lateral pipes. This has specific advantages of uniform water application, subsoil water application, minimum risk and low cost as compared to surface

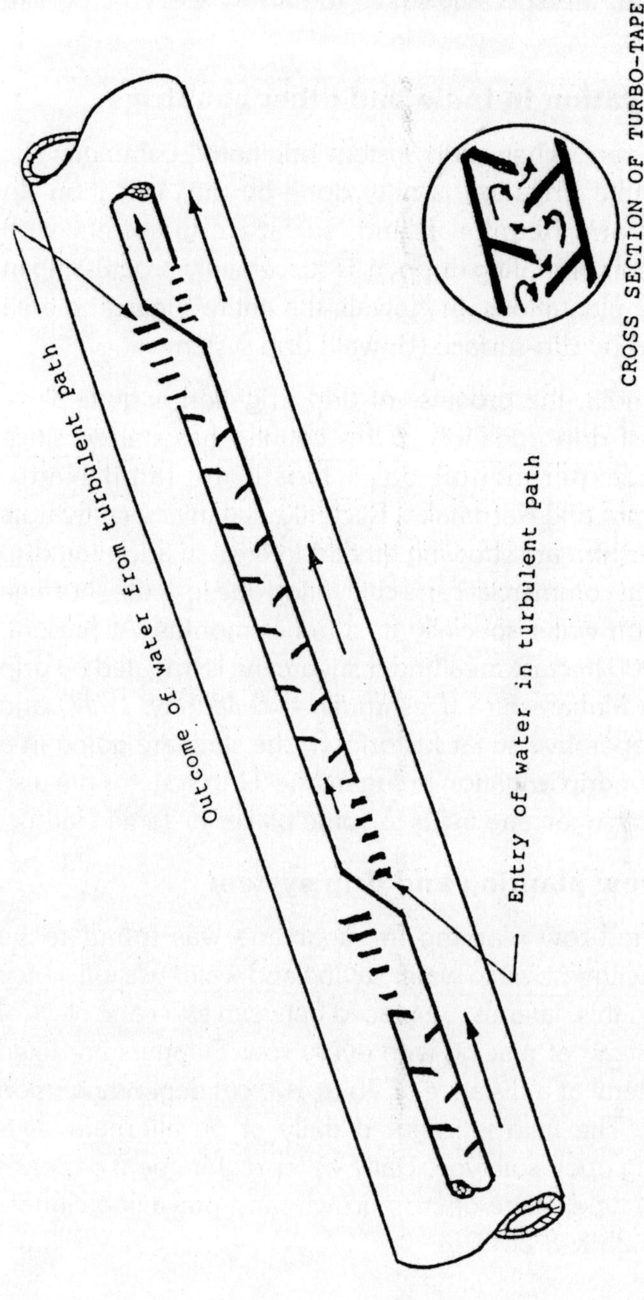

Fig. 8.7 : Turbo-tape

drip system. It is specially suited to narrow spaced crop like sugarcane.

Drip irrigation in India and other countries

In Israel, where drip system originated commercially, almost entire irrigation is now done by drip irrigation. In Mauritius and Reunion Island, surface drip system with pressure compensating drippers is successfully progressing in sugarcane plantations. In Hawaii, the entire sugarcane area is irrigated by sub-surface (Bi-wall) drip system.

In India, the progress of drip irrigation is quite slow. The use of drip irrigation in this country has started since 1980 on experimental basis mostly in Tamil Nadu, Maharashtra and Karnataka. Recently, sugarcane cultivators in Maharashtra are showing greater interest in adopting drip irrigation in commercial cane cultivation due to acute shortage of irrigation water specially in summer months. At present, nearly 4000 hectare area under sugarcane is irrigated by drip system in Maharashtra (Deshmukh and Jadhav, 1997) and many cooperative sugar factories in the state are going in a big way for drip irrigation in sugarcane. Drip systems are also installed in sugarcane fields in some places in Tamil Nadu.

Paired row planting and drip system

Paired-row planting in sugarcane was found to be superior with respect to water saving and water use efficiency (WUE). In this, laterals are placed between two cane rows of a pair instead of placing with every row. Drippers are fixed on the lateral at a distance of 75 to 100 cm depending upon soil type. The field is irrigated daily or on alternate days depending upon soil type. Daily water requirement depends upon soil type, stage of crop growth and prevailing climate (Tables 8.25 & 8.26).

Table 8.25 : Cane yield, water saving and water use efficiency (WUE) in sugarcane planted at two-row arrangements and irrigated daily or on alternate day with surface and sub-surface drip methods in clay soil at Pune, Maharashtra

Details of treatment			Cane yield (t/ha)	Cane yield increase (%)	Water applied (ha-cm)	Water saving (%)	WUE (t/ha-cm)
Irrigation method	Row arrangement	Irrigation					
Surface drip	Paired	Daily	175.5	36.9	107.6	50.1	1.63
Surface drip	Single	Alternate day	154.2	20.3	187.6	12.9	0.82
Sub-surface drip	Paired	Daily	143.8	12.2	93.9	56.4	1.53
Sub-surface drip	Paired	Alternate day	142.5	11.9	93.9	56.4	1.52
Furrow	Single	Conventional	128.2	-	215.4	-	0.60

Source: Hapse et al.,1994

8.7 **Sugarcane fertigation**

Fertigation involves application of liquid or solid soluble fertilizers through drip irrigation at slow and controlled rate at the root zone of the crop. Due to controlled application of fertilizers the losses such as leaching, volatilization, denitrification etc. are avoided.

Table 8.26: Cane yield, water saving and WUE in sugarcane planted at two-row arrangements and irrigated with drip and furrow irrigation methods in silty clay and clay soils at five locations of Nasik and Solapur districts of Maharashtra

Details of treatment		Cane yield (t/ha)	Cane yield increase (%)	*Water used (ha-cm)	Water saving (%)	WUE (t/ha-cm)
Irrigation method	Row arrangement					
Surface drip	Paired (150:75 cm)	126.8	27.4	159.8	37.4	0.793
Surface drip	Single (100:100 cm)	118.4	19.0	191.5	25.0	0.618
Sub-surface drip	Paired (150:75 cm)	125.3	25.9	152.5	40.2	0.822
Sub-surface drip	Single (100:100 cm)	111.7	12.3	172.8	32.3	0.646
Furrow	Paired (150:75 cm)	101.2	1.7	226.7	11.2	0.446
Furrow	Single (100:100 cm)	99.5	-	255.2	-	0.390
CD (0.05)		10.7	-		-	-

Source: Deshmukh and Jadhav, 1977

* Including rainfall

Advantages of fertigation are:

i) Higher fertilizer use efficiency along with increased water use efficiency

ii) Application of fertilizers at different crop stages as per crop demand is possible

iii) Nutrients are available at higher rate as soil moisture is maintained at field capacity

In a study conducted by Shinde *et al.* (2000) at Rahuri on clay loam soil, application of solid or liquid fertilizers as fertigation proved significantly better than conventional method of surface applied solid fertilizers (Table 8.27). Application of liquid fertilizers through drip was found more efficient as it saved 25% fertilizers as well as increased the cane yield by 20% compared to conventional fertilizer application. Kadam (1994) and Shinde *at al.* (1992) also indicated the possibility of saving 25% of nitrogen by applying it through drip.

8.8 **Irrigation and fertilizer use**

The uptake of nutrients from the soil is conditioned by the soil moisture. The mobility of the nutrients in the soil is a major consideration which is influenced by soil moisture content. In the immediate vicinity of the roots only a small portion of the total nutrient requirement of the plant is available. Nutrient translocation from other soil areas to the plant roots, therefore, plays an important role. It occurs mainly by mass flow (with the water moving to the plant roots to compensate for evaporation losses) and diffusion (movement to level up concentration differences in the soil solution). Thus, moisture stress acts as a limiting factor in nutrient uptake, resulting in decreased yield of cane and sugar. Full advantage

Table 8.27: Yield of *suru* sugarcane and its ratoon as influenced by levels and methods of fertilizer application at Rahuri, Maharashtra

Fertilizer application	Cane yield (t/ha)	
	Plant	Ratoon
*Conventional	134.0	123.9
RDSF (NTD) through Drip	158.8	152.5
125% RDLF through Drip	170.7	167.6
100% RDLF through Drip	166.3	155.2
75% RDLF through Drip	161.3	149.8
50% RDLF through Drip	140.5	127.3
CD (0.05)	16.9	11.0

Source: Shinde *et al.*,2000

(Recommended dose of fertilizer was 250, 125 and 125 kg N, P_2O_5 and K_2O/ha, respectively)

*Conventional = Whole amount of P & K was applied at planting and N in four splits (10% at planting + 40% 8 WAP + 40% at earthing up)

RDSF = Recommended dose of solid fertilizer

RDLF = Recommended dose of liquid fertilizers

NTD = Only nitrogen through drip

Fertigation was scheduled at an interval of 15 days started from one month after planting in 10 equal splits.

In surface irrigated treatments the irrigations were scheduled at 75 mm CPE with 70 mm depth. In drip, the irrigation was scheduled at every alternate day.

through fertilizers can be availed by providing adequate moisture to the crop.

In a study there was progressive increase in yield (Table 8.28) with increase in the number of irrigations. The same was not true with nitrogen. Yield increased from no nitrogen to 150 kg N/ha but further additional dose of N resulted in an significant increase in yield only under the higher irrigation levels. Therefore, increased level of N fertilization had to be associated with increased frequencies of irrigation to realize higher cane and sugar yields. Under low soil moisture availability it is necessary to limit the nitrogen application rate because higher N dose does not promote more growth than the available soil moisture can sustain. On the other hand, the rate of N application under high soil moisture conditions, should be adjusted to enable the plant to make full and efficient use of favourable soil moisture to express its full yield potential.

8.9 Sugarcane production under inadequate irrigation facilities

A sizeable area under sugarcane in the country, though reported as irrigated, does not receive optimum quantity of water for irrigation. Lack of dependable irrigation source has also been responsible for this situation. In areas where crop is grown with canal irrigation, the possibility of getting adequate irrigation is limited during summer months due to closure of canal. The irrigation is often limited as the facilities for supplementing with lift irrigation is also not to the desired level. However, the loss of sugarcane yield due to limited irrigations may be minimised by selecting suitable varieties and resorting to certain agronomic practices.

i) Selection of suitable varieties

ii) Trash mulching

iii) Skip-furrow irrigation

iv) Trench planting

v) Applying irrigation during most critical period of crop water need

vi) Use of moisture absorbants

Table 8.28: Interrelation of nitrogen and moisture levels in influencing cane yield (t/ha) in sandy loam soil at Hyderabad, Andhra Pradesh

N level (kg/ha)	Irrigation at available soil moisture depletion of			
	25%	50%	75%	Mean
0	72.1	70.6	66.4	70.0
50	92.6	83.7	73.5	83.2
100	102.2	87.8	78.5	89.8
150	110.0	95.4	84.9	96.8
200	118.7	98.7	87.5	100.6
250	127.1	103.3	90.6	107.0
300	134.6	106.0	91.6	110.9
Mean	108.3	92.2	81.9	-

CD (0.05) Nitrogen levels (N) = 5.4 Irrigation regimes (I) 3.4

NxI = (a) The difference between the same N level at different moisture regimes = 9.1

(b) The difference between the same moisture regimes at different N levels = 6.5

Source: Reddy *et al.*, 1978

i) Selection of suitable varieties: There are varieties which are able to withstand the limitation of water to a higher extent. The margin of drop in yield in such varieties under lower moisture regimes remains low and they yield comparatively higher under limited supply of water. Varieties Co 671, Co 740 and Co 6304 at Coimbatore, Co 6407 at Parvaranagar, Co 1148 at Lucknow and Co 1148 and CoJ 67 at Jalandhar were found better to adopt themselves to drier moisture regimes (Tables 8.29 to 8.33).

Table 8.29: Yield (t/ha) performance of different varieties of sugarcane irrigated at different levels of available soil moisture depletion (ASMD) at Coimbatore, Tamil Nadu (Mean of data from 1988-89 and 1989-90)

Variety	Irrigation at ASMD of			
	25%	50%	75%	Mean
Early				
CoC 671	107.2	100.8	94.0	99.6
Co 7201	115.9	99.5	76.7	98.5
Co 6806	90.7	87.2	68.2	82.2
Mid late				
Co 6304	125.5	96.3	92.7	104.1
Co 1148	120.4	105.0	83.9	101.5
Co 419	117.0	104.0	84.9	99.9

Source: Jambulingam *et al.*,1999

Table 8.30: Yield (t/ha) performance of different sugarcane varieties irrigated at various levels of ASMD (March-June) at Coimbatore

Variety	*Irrigation at ASMD of			
	40%	60%	80%	Mean
Co 740	124.1	116.6	92.1	111.0
Co 419	114.7	97.3	91.4	101.2
CoS 510	84.3	80.5	71.5	78.8
Co 853	81.5	83.0	71.1	78.6
Co 997	82.4	82.0	67.6	77.3
Co 1148	78.4	69.1	71.7	73.1
Co 527	68.4	64.7	55.9	63.0
Co 658	76.7	71.9	48.9	65.8
Co 1158	66.7	59.8	49.9	58.8
Mean	86.4	80.6	68.9	-

CD (0.05): Varieties (V) = 7.4, Irrigation (I) = 9.8, VxI = NS

(*Irrigation at 40%, 60% and 80% ASMD were corresponding to 6-8, 10-12 and 20-25 days' intervals, respectively)

Source: Singh and Reddy, 1980

Table 8.31: Yield (t/ha) performance of different sugarcane varieties under various irrigation intervals in black soil at Parvaranagar, Maharashtra.

Variety	Irrigation interval (days)			
	20	30	40	Mean
Co 6407	125.5	96.8	96.6	106.3
Co 740	106.8	93.1	81.0	93.3
Co 7219	97.5	92.5	76.0	88.6
Co 8081	95.8	91.9	75.5	87.8
Co 7704	95.2	89.6	74.1	86.3
Mean	104.1	92.8	80.7	-

CD (0.05): Varieties (V)=2.8, Irrigation (I)=1.5, VxI=NS

Source: Kadu *et al*, 1988

Table 8.32: Yield (t/ha) performance of different sugarcane varieties irrigated at various levels of ASMD during summer at Lucknow, Uttar Pradesh

Variety	Irrigated at ASMD of			
	25%	50%	75%	Mean
Co 1148	63.7	68.3	51.6	61.2
Co 1158	76.4	51.6	45.4	57.8
CoJ 64	73.9	61.4	42.4	59.2
Mean	71.3	60.4	46.5	-

SE: Variety(V) and Irrigation(I)=0.65, VxI=0.87

Source Yadav and Prasad, 1988

Table 8.33: Yield (t/ha) performance of different sugarcane varieties irrigated at various moisture regimes in sandy loam soil at Jalandhar (Punjab) (Mean of data from 1982-83 and 1983-84)

Variety	Irrigated at IW/CPE ratio of			
	0.9	0.7	0.5	Mean
Co 1148	76.5	71.6	68.0	72.0
CoJ 67	74.0	70.5	66.2	70.3
CoJ 64	69.4	65.4	60.9	65.3
Mean	73.4	69.2	65.0	-

CD (0.05): Varieties(V)=3.39, Irrigation(I)=2.58, VxI=5.33

Source: Chaudhary and Kanwar, 1986

ii) Trash mulching: The use of mulches is a most practical measure to increase the effectiveness of irrigation by reducing evaporation loss of moisture from soil surface (Tables 8.34 & 8.35). Trash mulching keeps the soil moisture at a higher level for a longer time as compared to uncovered soil surface. The trash cover of 10 cm thickness should be spread in inter-row spaces after completion of germination in plant cane and in ratoon crop at the time of its initiation.

iii) Skip–furrow irrigation: Economy in irrigation water upto 30-40% may be made by way of irrigating alternate furrow instead of every furrow without sacrificing cane yield (Table 8.36)

iv) Trench planting: To make better use of water per irrigation, the planting should be done in trenches. Trench planting encourages deep rooting in cane plants.

Table 8.34: Cane yield (t/ha) as affected by trash mulching under different levels of ASMD at Lucknow, Uttar Pradesh

Irrigation at	Mulching		
	No mulching	Mulching	Mean
75% ASMD	63.8 (3)	76.4 (2)	70.1
50% ASMD	69.3 (4)	82.6 (3)	76.0
Mean	66.6	79.5	-

(Figures in parentheses indicate number of irrigations applied during tillering phase in summer)

Source: Motiwale and Singh, 1977

v) Applying irrigation during most critical period of crop water need: Ensuring irrigation at critical period of water need of the crop and deferring the same at somewhat less critical period is beneficial where there is limited water supply.

During formative phase one irrigation in April and one in May was found to be absolutely necessary for good growth of the plants at Shahjahanpur. The results (Table 8.37) obtained at Lucknow indicated that in case of single irrigation, applying it on April 15 was highly yield productive followed by May 10. In case of two irrigations, April 15 and May 10 was significantly superior than other combinations and was at par with three irrigations given on April 15, May 10 and June 5. Irrigation on June 5 was the least productive. (Singh

Table 8.35: Cane yield as affected by trash mulching at different places in India. (Crop was irrigated by alternate-furrow irrigation)

Treatment	Cane yield (t/ha)		Soil type	Reference
	1988-89	1989-90		
Jalandhar, Punjab				
No mulching	32.6	35.8	Sandy	Kanwar
Mulching	59.2	57.9	loam	*et al.*,1992
CD (0.05)	9.7	6.3		
Karnal (Haryana)				
No mulching	48.5	47.2	Sandy	Kumar &
Mulching	64.2	57.2	loam	Srivastava,
CD (0.05)	12.6	13.3		1991
Muzaffarnagar (Uttar Pradesh)				
No mulching	60.4	53.3	Sandy	Malik
Mulching	76.6	51.9	loam	*et al.*,1996
CD (0.05)	1.5	NS		
Sehore (Madhya Pradesh)				
No mulching	70.6	55.8	Clay	Sharma
Mulching	80.7	80.2	loam	& Verma,
CD (0.05)	8.0	5.7		1996
Cuddalore (Tamil Nadu)				
No mulching	102.0	127.9	Clay	Pandian
Mulching	116.4	135.5	loam	*et al.*,1992
CD (0.05)	14.3	7.0		

Table 8.36: Cane yield (t/ha) as affected by different modes of furrow irrigation and trash mulching at Belgaon, Karnataka

Irrigating	Mulching		
	No mulching	Trash mulching	Mean
Every-furrow	97.4	110.5	104.0
Alternate-furrow	105.1	114.5	109.8
Mean	101.3	112.5	-

Source: Hunsigi and Parmeshwarappa,1975

et al., 1976). Similarly, at Pantnagar, two irrigations applied on 45th and 75th days of planting produced the maximum cane yield (Singh *et al.*, 1986). In case of one irrigation, giving it early in the season proved to be more effective (Table 8.38).

Bose and Thakur (1977) at Pusa observed that even in case of single irrigation, the critical period lay between 45 and 60 days of planting, At Anakapalle, when irrigation water is available to give only 2 or 3 irrigations, one irrigation should necessarily be given in May (Prasada Rao, 1983).

All these studies indicate that during pre-*monsoon* period, early part of the formative phase of the crop is most critical for water need.

During post-*monsoon* period, one irrigation applied by the end of November gave best results. If harvesting is to be delayed, crop responds well to one more irrigation applied by the end of March at Shahjahanpur (Anonymous, 1941).

Table 8.37: Yield of sugarcane in relation to irrigation applied on different dates during pre-*monsoon* period in sandy loam soil at Lucknow, Uttar Pradesh (Crop planted in second week of February, Cv.Co 1158)

Irrigation applied on	Cane yield (t/ha) 1970-71	1971-72	Mean
One irrigation			
April 15	54.4	29.6	42.1
May 10	45.7	27.2	36.5
June 5	27.2	25.1	26.2
Two irrigations			
April 15 & May 10	58.9	45.9	52.4
April 15 & June 5	47.6	32.5	40.0
May 10 & June 5	48.7	30.6	39.7
Three irrigations			
April 15, May 10 & June 5	64.1	47.5	55.8
CD (0.05)	6.2	2.1	

Source: Singh *et al.*,1976

vi) Use of moisture absorbants: Superabsorbants like *Jalshakti* have the unique property of absorbing water many times its weight and making it available to the plants subsequently resulting in overcoming moisture stress. Field studies conducted at different places in India indicated beneficial effect of using *Jalshakti* at only few places (Table 8.39). Thus the situation under which beneficial effect is expected needs to be assessed.

Table 8.38: Yield of sugarcane in relation to irrigation applied after different days of planting in clay loam soil at Pantnagar (Crop planted on March 5, 1975 and Feb 20, 1976, Cv.Co 1158)

Irrigation (DAP)	Cane yield (t/ha)		
	1975-76	1976-77	Mean
75	69.6	62.7	66.2
90	64.8	58.0	61.4
45&75	76.0	70.0	73.1
Mean	70.1	63.6	-
CD(0.05)	4.2	4.3	

Source: Singh *et al.*, 1986

8.10 **Soil moisture and frost**

Occurrence of frost in northern India is a common feature during the winter months (December – January). Frost affects the sugarcane crop in three ways. Firstly, the tips of leaves are killed and guard cells of stomata are rendered inactive over the entire surface of the leaves; secondly, the buds and leaf primordia are injured; and thirdly, there occurs deterioration of the juice quality. Singh and Singh (1974) suggested keeping soil at higher moisture level during freezing to protect the sugarcane crop against injury caused by frost.

8.11 **Water quality**

The concentration and composition of dissolved constituents in water determine its quality for irrigational use. The quality of water is a factor of immense importance in the context of sugarcane production. In general, water containing

Table 8.39: Cane yield as affected by use of *Jalshakti* (20 kg/ ha) at different places in India (Crop irrigated by furrow method)

Use of Superab-sorbant	Cane yield (t/ha)		Soil type	Reference
	1988-89	1989-90		
	Jalandhar (Punjab)			
No *Jalshakti*	47.6	45.3	Sandy	Kanwar
Jalshakti	55.5	51.6	loam	*et. al.*,
CD (0.05)	NS	6.3		1992
	Karnal (Haryana)			
No *Jalshakti*	68.7	65.2	Sandy	Kumar &
Jalshakti	68.0	64.3	loam	Srivastva,
CD (0.05)	NS	NS		1991
	Muzaffarnagar (Uttar Pradesh)			
No *Jalshakti*	65.5	50.0	Sandy	Malik
Jalshakti	74.3	45.3	loam	*et al.*,
CD (0.05)	1.5	NS		1996
	Sehore (Madhya Pradesh)			
No *Jalshakti*	81.7	80.6	Clay	Sharma
Jalshakti	85.9	83.2	loam	&Verma,
CD (0.05)	NS	NS		1996
	Cuddalore (Tamil Nadu)			
No *Jalshakti*	116.4	135.8	Clay	Pandian
Jalshakti	122.1	122.6	loam	*et al.*,
CD (0.05)	NS	NS		1992

higher soluble salts (electrical conductivity >2.0 mmhos/cm) is considered to be of poor quality for sugarcane. Poor quality waters are dominated by cations like sodium and magnesium and anions like chlorides, sulphates and bicarbonates. Growth, yield and quality of sugarcane are affected adversely by poor quality water. Besides, continuous use of such water leads to development of soil salinity and sodicity. The condition of sodicity is created when there is predominance of sodium ions in irrigation water.

Salinity of irrigation water is one of the serious problems affecting sugarcane production in several parts of the country. About 9 lakh hectares of sugarcane area in India is reported to be affected, to varying degrees, by this problem (Sundara 1996). In many sugar factory areas under arid and semi-arid tracts, groundwater is becoming saline because of the depleting groundwater potential due to over exploitation. Broadly speaking, the factors affecting the quality of groundwater resources of the country can be divided into three main groups:

i) Water of arid and semi-arid regions having rainfall below 45 cm per annum,

ii) Water quality as influenced by hydrological conditions as high water table in irrigated areas and river basins, and

iii) Water quality of wells in some areas of coastal regions as influenced by sea water intrusion and inundation.

Of all these factors, aridity is the most important single factor responsible for very high salinity of well water.

Sugarcane yield reductions ranging from 10-50 per cent due to saline water irrigation have been reported by Ramalingaswami *et al.* (1986) and Kumaraswamy (1989).

The growth (Table 8.40) of cane and uptake of major nutrients tended to decrease with increase in electrical conductivity (EC) of water. Appreciable reduction in cane yield and depression in cane juice quality were noted when EC of irrigation water was more than 2.0 mmhos/cm due to higher salt content. Continuous application of poor quality water was noted to have increased pH, EC and exchangeable sodium of soil (Table 8.41).

Adverse effects of saline irrigation water can be minimized and productivity improved by selecting salt tolerant varieties such as CoC 671, Co 7201, Co 6806, Co 8145, Co 8208 & Co 7717 (Table 8.42) and adopting certain agronomic practices like trench or pit planting (Table 8.43), and applying organics like FYM or press mud combined with trash mulching (Table 8.44).

Table 8.40: Effect of quality of irrigation water on growth, yield and quality of sugarcane at Anakapalle, Andhra Pradesh

EC(mmhos/cm) of irrigation water	Cane height (cm)	Stalk population per pot	Yield (kg/pot)	Sucrose (%) juice	Purity (%) juice
Control	242.9	4.96	3.84	18.25	88.3
2.0	233.5	4.51	3.31	17.42	85.5
4.0	221.9	3.65	2.88	16.60	82.7

Source: Ramalingaswami *et al.*, 1986

Table 8.41: Effect of quality of irrigation water on soil pH, EC and ESP as recorded on different days after planting sugarcane at Anakapalle, A.P.

EC (mmhos/cm) of Irrigation water	Initial	Days after planting		
		120	210	At harvest
Soil pH				
Control	7.55	7.56	7.57	7.59
2.0	7.55	7.65	7.82	7.91
4.0	7.55	7.76	8.17	8.28
Electrical conductivity(EC)				
Control	0.368	0.332	0.333	0.361
2.0	0.368	0.341	0.368	0.425
4.0	0.368	0.358	0.428	0.539
Exchangeable sodium per cent (ESP)				
Control	4.62	4.63	4.50	4.42
2.0	4.62	4.87	6.20	7.36
4.0	4.62	5.30	7.50	9.29

Source: Ramalingaswami *et al.*, 1986

Table 8.42: Yield and quality of sugarcane varieties as affected by saline water irrigation at Coimbatore, Tamil Nadu

Variety	1991-92			1993-94		
	Cane yield (t/ha)	Sugar yield (t/ha)	CCS (%)	Cane yield (t/ha)	Sugar yield (t/ha)	CCS (%)
CoC 671	84.8	12.29	14.49	68.2	8.47	12.3
Co 7201	83.6	8.97	10.73	-	-	-
Co 6806	80.1	9.87	12.32	60.2	6.81	11.24
Co 8145	-	-	-	76.4	8.28	10.83
Co 7717	-	-	-	76.2	7.04	9.10
Co 775	74.1	8.49	11.46	-	-	-
Co 8208	60.6	7.66	12.65	73.7	8.51	11.62
Co 7704	55.1	6.88	12.49	29.0	3.08	9.93
CD(0.05)	9.8	-	0.52	12.0	1.42	1.02

Source: Sundara and Reddy, 1994 and Sundara, 1996

Soil – Sandy loam (alfisol), pH 7.8, EC 0.6 mmhos/cm

Irrigation water- pH 7.6, EC 3.0-3.5 mmhos/cm, salts (mg/l) present were: Carbonates 1.4, Bicarbonates 6.0, Sulphates 15.5, Chlorides 23.0, Magnesium 13.6, Sodium 6.1, Calcium 23.0 and Potassium 0.37 and SAR 1.85

Table 8.43: Effect of methods of planting on cane yield and soil pH and EC under saline water irrigation at Coimbatore, Tamil Nadu

Planting method	1991-92	1993-94		
	Cane yield (t/ha)	Cane yield (t/ha)	*Soil pH	*Soil EC(mmhos/cm)
Pit	81.9	-	-	-
Trench	77.2	73.3	8.1	0.44
Ridge and furrow	65.4	54.6	8.4	0.57
CD (0.05)	10.2	15.9	-	-

Source: Sundara and Reddy, 1994 and Sundara, 1996

(Note: Experimental soil and irrigation water characteristics as in table 8.42)

*At 210 days after planting

8.12 **Drought and sugarcane**

Moisture stress is one of the serious problems of sugarcane production throughout India. Nearly 70 per cent of the sugarcane area in subtropical and 65% per cent in tropical India experiences moisture stress at one or the other stage of crop growth (Singh, 1997). However, the intensity of the problem in specific region is different. In most of the areas, moisture stress occurs from March to June (summer season) during germination and tillering or tillering and early cane formation (formative phase) phases. During this period, crop

Table 8.44: Effect of different organics on cane and sugar yields and soil pH and EC under saline water irrigation at Coimbatore, Tamil Nadu

Organics	Yield (t/ha)		Soil (210 DAP*)	
	Cane	Sugar	pH	EC (mmhos /cm)
Control	56.4	6.91	8.4	0.55
FYM (24 t/ha)	63.7	7.87	8.1	0.50
Trash mulch(TM)	60.1	7.29	8.3	0.54
FYM+TM	73.5	9.02	8.1	0.47
Press mud(25t/ha)	61.2	7.65	8.1	0.47
Press mud +TM	63.3	8.22	7.9	0.47
CD (0.05)	13.0	0.94	-	-

Source: Sundara,1996

*DAP = Days after planting

(Note: Experimental soil and irrigation water characteristics as in table 8.42

faces both soil and atmospheric droughts[1] because of high temperature, bright sunshine, hot winds and low humidity.

[1] *Drought*: It is a meteorological term that refers to only substantial period in which rainfall fails to keep up with potential evaporation. In agricultural context, when evaporation exceeds precipitation and the atmosphere and edaphic moistures fail to sustain plant growth, the conditions are known as drought.

Soil drought: It occurs when soil moisture is not adequate to keep pace with moisture absorption and transpiration through leaves.

Atmospheric drought: It occurs when transpiration of moisture through leaves is in excess of moisture absorption due to higher atmospheric temperatures, extremely low atmospheric humidity, bright sunshine and hot winds. It may occur even under conditions of relatively higher soil moisture due to excessive increase in transpiration which exceeds the rate of absorption of water from the soil.

Growth of the cane which suffered during formative phase due to moisture stress could not be made up even when soil moisture become adequate during elongation phase. Even during rainy season, crop often experiences water deficit in areas having low rainfall or cyclic drought caused due to failure of *monsoon*. Such drought continues to be a challenge due to the unpredictability of its occurrence, severity, time and duration. Denial of rainfall, in spite of providing adequate irrigation during July-August in rainy season in grand growth phase depressed the cane yields to the tune of 33.2 per cent at Anakapalle, Andhra Pradesh (Prasada Rao, 1983). Cane yield comparison has also been made between normal year (1986) and drought year (1987) at Lucknow by Yadav (1989). He reported that rain-fed cane suffered 48 per cent losses due to rainy season drought and the yield of irrigated sugarcane was also reduced by 23 per cent. It clearly indicates that when climatic conditions are not favourable during the grand growth phase, due to drought caused by failure of south-west *monsoon*, the cane yields are adversely affected even when the crop is irrigated.

Effect of summer season drought on sugarcane crop can be alleviated to a greater extent by irrigations. For irrigation purposes, canals, wells and tanks are the chief sources of water. In summer, from March to June, most of the rivers, tanks and wells dry up and scarcity of water arises. This leads to shortage of water for sugarcane thus affecting the yields.

In the same climate, difference in soil conditions (texture, depth and salt contents) modify the drought proneness. Severity of drought increases in light a textured soils. The moisture holding/storage capacity of a sandy soil is much less than loam or clay soils. Evaporation, which is very prominent during drought, has been observed to be proportional to the rate at which water moves to the surface from inner capillaries. Sandy soils are at a disadvantage, because such capillary movement is faster in light textured soils.

The depth of the soil determines its total capacity for moisture storage, and sugarcane in a shallow soil would be more severely affected by drought than that in a deep one, because the latter would act as a sizeable reservoir of soil water. Under drought prone conditions selection of a deep soil would be insurance against the ravages of drought.

The saline soil, which contains appreciable amounts of salts, produces even greater ill effects as drought results in a very high concentration of salts is the rhizosphere. In saline soils, there is increase in osmotic pressure of the soil solution. Increase in osmotic pressure would contribute to increase in overall soil moisture stress and make it difficult for the plant to absorb even the small amounts of water available in the root zone. Hence, the drought effect will be aggravated in soils with light texture, shallow depth and high salt contents.

8.12.1 Plant symptoms of drought: Morphological indices such as alterations in the colour of leaves, noticeable drooping of leaves, loss in turgidity are some of the prominent symptoms of drought effects. However, the nature of injury caused by soil drought is different than atmospheric drought in some respects. The soil drought first affects the leaves which starts from the lowest to the upper leaves. Generally, the desiccation starts from the tips and then proceeds towards base along with the margins. The fresh green colour of the leaves changes to more bluish and ultimately leaves become yellow and subsequently dry up. The yellowness caused by water deficiency is quite different than that of nitrogen deficiency. The latter shows a pale yellow colour whereas the former exhibits dark bluish yellow. The worst affected plants show wilting by curling of the leaves whose margins turn inwards. The extent of curling differs from variety to variety.

In case of atmospheric drought the upper top most first and second emerged leaves are affected whereas lower leaves remain normal greenish. The drying of leaves starts from the tip and spreads very quickly all along the lamina towards the base. Such type of drought generally occurs during last week of May to second week of June in subtropical India when weather extremes prevail due to extremely high atmospheric temperature (temperature crossing 40^0 C), low humidity (10-38%), bright sunshine and hot winds.

8.12.2 **Adaptation to water stress**: Many mechanisms exist whereby several sugarcane cultivars may cope with the consequences of drought period. In order to maintain reasonably higher water content under dry conditions, plants may either reduce the transpiring surface or increase the absorbing surface. Root development and water uptake, transpiration and micro- and macro-morphological adaptation, osmotic adjustment have important consequences on drought adaptation.

Under dry conditions there are various anatomical adaptation of the roots such as cell walls of the parenchyma in the roots become thicker due to deposition of suberin and lignin.

If plant does not obtain enough water to balance unavoidable loss, it often reduces its transpirational loss of water by reducing the number and size of leaves, curling of leaves and closure of stomata. The reduction in number and size of leaves has been ascribed to a highly effective means of adaptation to occasional or periodic drought. Under moisture stress conditions, reduction in evaporating surface through curling of leaves is common in some varieties. This enables the plants to reduce water losses, possibly because of higher number of bulliform cells. Curling and inward rolling of leaves are known to reduce the radiation load on leaves, thereby reducing the leaf temperature which in turn decreases water loss. Some varieties like Co 1148, Co 6806, and BO 43 start curling at moderate stress while other do not curl even under severe stress.

Grantz and Meinzer (1988) made exhaustive studies on the mechanism and consequences of stomatal response to drought in field grown sugarcane. The stomatal conductance fell rapidly with the onset of drought but exhibited partial recovery following resumption of irrigation. During the drought period, soil water content fell from about 24% to 14% (by volume) while the rapid and sensitive stomatal response protected the leaf water status and prevented it from any substantial fall. The continuing fall in soil water content was closely paralleled by the fall in stomatal conductance. They also suggested that the signals transmitted from root to shoot might communicate information regarding the soil environment. The stomatal response to the soil environment might have an important part to play in protection of the shoot from water deficiency that could damage the photosynthetic apparatus.

Rutherford (1989) investigated the assessment of proline accumulation as a mechanism of drought resistance. Proline is commonly accumulated by stressed plants and may be of benefit as a reservoir of nitrogen and energy for recovery after alleviation of stress and may prevent denaturation of proteins essential for cell activity. In a pot culture study conducted by Chiranjivi Rao and Asokan (1978) involving 22 sugarcane varieties, accumulation of large quantities of free proline (up to 1600% increase) was observed when plants were under water stress. Drought resistant varieties seem to accumulate comparatively higher proline than the susceptible ones. Thus, an estimation of free proline could serve as a valuable indicator of drought resistance in cane varieties.

8.12.3 **Effect of moisture stress on sugarcane crop**: The response of crop to drought may vary greatly depending on variety and stress severity. These responses include changes in plant growth, yield, quality and metabolic activities.

i) Reduced growth and yield: Plant (roots, stalks and leaves) growth is reduced severely. In general, moisture stress reduces leaf area, number of tillers, millable canes and internodes, cane length and girth which finally results in reduced cane yields(Table 8.45). Drying of leaves, tiller mortality and stunted growth are the common features of moisture stress affected sugarcane.

ii) Decreased photosynthesis: The rate of photosynthesis is affected by water stress directly as a result of tissue dehydration and indirectly as a result of stomatal closure and its influence on gas exchange. The stomata which involve both in transpiration and photosynthesis when open to transpire, more carbon dioxide from atmosphere penetrate into the plants for photosynthesis. Both processes cease when the stomata are closed under moisture stress conditions. Moreover, photosynthesis depends largely on relative water content of leaves, which in turn is influenced by soil moisture content. Therefore, to maintain high photosynthetic rates over an extended period of time, it is essential to provide enough water to leaves. However, sugarcane genotypes which can maintain leaf water potential and absorb more nutrients may resist decrease in carbon assimilation during stress and hence yield more.

iii) Reduced quality: Drought has deleterious effects on the sugarcane quality particularly brix, sucrose and purity of juice (Table 8.46) . This deleterious effect is greater in late maturing varieties than the earlier ones. The magnitude of effects differ with severity of moisture stress. Moisture stress also delays maturity which is more in late than early maturing varieties. In tropical India, depression in juice quality is also related to greater production of immature internodes following the release of moisture stress. In tropics, therefore, harvesting of such crop late in the season has been suggested to get maximum yield.

Ahmad and Khan (1991) studied the effect of drought on quality of sugarcane and reported that the brix, sucrose and purity of juice and sugar recovery per cent of cane were lower in 1987-88 (drought year) in seven sugar factory areas in the *tarai* region of Uttar Pradesh than 1986-87 (normal rainfall)

iv) Increased pests: It has often been seen that the ratoon stunting disease (virus) and wilt (fungus) infestation is much more under conditions of moisture stress than when sufficient moisture is available, more so, if it is a ratoon crop.

Table 8.45: Effect of soil moisture stress during formative phase on cane juice quality and sugar yield at Rudrur, Nizamabad, Andhra Pradesh

Treatment	Sucrose (%)	Purity (%)	CCS (t/ha)
Normal irrigation	15.6	88	13.5
Soil moisture stress	12.4	77	7.7

Source: Rao Mukunda *et al.*,2000

Table 8.46: Commercial cane sugar per cent in sugarcane grown under irrigated, semiirrigated and rainfed conditions at Baramba, Orissa

Variety	Moisture supply			Mean
	Irrigated	Semi-irrigated	Rainfed	
		December, 1996		
Co 740	9.08	8.36	7.36	8.23
Co 8021	9.13	8.38	7.74	8.42
Mean	9.10	8.37	7.55	-
		January, 1997		
Co 740	9.85	9.11	8.15	9.04
Co 8021	9.95	9.28	8.30	9.18
Mean	9.90	9.20	8.22	-

Source: Rajkumar and Khuntia, 1998

Table 8.47: Drought resistant parameters at 2-3 leaf stage in different sugarcane varieties (Pot experiment)

Variety	Shoot: root ratio	Cuticular deposition (lower epidermis)	Length of main vertical roots (cm)	Nature of roots	PWP	DR/DS *
CoS 510	1.5:1	More	34	Well branched	3.12	DR
Co 312	2.0:1	More	32	Well branched	3.25	DR
Co 859	3.5:1	Much less	21	Poorly branched	3.60	DS
Co 453	4.0:1	Less	28	Moderately branched	3.62	DS
CoS 416	5.0:1	Medium	24	Moderately branched	4.45	DS

Source: Singh, 1964

PWP (Permanent wilting point) for sunflower was 6.4 in this soil

DR=Drought resistant, DS=Drought susceptible

8.12.14 **Characteristics of drought tolerant sugarcane varieties**

- Drought tolerant sugarcane varieties (DTSV) are generally characterized by a deeper and extensive root system (Table 8.47) to tap the water from deeper horizons. Sugarcane variety Co 740 was found drought resistant in a study at Kolhapur because of its deep root penetration (Magdum and Divekar, 1962). The roots of DTSV also remain functional for long period under drought conditions. The capacity of Co 997 to resist drought was due to production of more functional roots in spite of prolonged drought (Lakshmikantham *et al.*, 1971).

- DTSV are capable of utilizing moisture from a level far below the wilting coefficient (Singh and Singh, 1974).
- DTSV have short narrow leaves with lesser leaf area per plant. There are also cuticular deposits on the epidermal cells of the leaves of these varieties (Table 8.47).
- DTSV have lower shoot:root ratio (Table 8.47). Ratio of total leaf surface to total absorbing surface appears to be a valuable criterion of drought resistance.
- Varieties showing higher stomatal diffusion resistance (SDR) and leaf water potential (LWP) are tolerant to soil moisture stress (Srivastava *et al.*, 1997).
- There is close relationship between drought tolerance and size, distribution and movement of stomata. There are large differences in stomatal sensitivity, size and population among sugarcane cultivars. Stomata in DTSV are sensitive to drought. Greater sensitivity of the stomata of Co 285 may be associated with its better drought resistance (Naidu and Bhagyalakshmi, 1973). Studies on stomatal movement indicated that the DTSV close their stomata earlier under stress conditions than the susceptible ones to control loss of water effectively and also open them quickly upon rewatering. Drought tolerant varieties, Co 285 and Co 1148 were found to have greater stomatal resistance (Venkataraman *et al.*,

1986). Varieties possessing low frequency and smaller stomata are more tolerant to drought condition. Cultivars, Co 1148 and BO 54 possessed relatively lesser number of small size stomata and hence were tolerant to drought (Singh and Ali, 1981).

- Bound water content at a stage just prior to permanent wilting level showed a wide range of variation in sugarcane varieties. DTSV have high bound water*
- Drought resistant varieties lose the turgidity earlier than susceptible types. Naidu and Bhagyalakshmi (1973) studied the relative turgidity and stomatal movement daily during drought cycles in four sugarcane varieties known for differences in their resistance to drought. Relative turgidity of drought resistant varieties Co 285 and Co 312 decreased rapidly to as low as nearly 60% at the end of drought. Nearly 90% of the stomata closed on the fifth day of drought. In drought susceptible varieties, the relative turgidity decreased less rapidly and reached 76% in Badilla and 80% in Kaludai Boothan, towards the end of drought. Only about 50% of the stomata closed in Kaludai Boothan on the seventh day.
- DTSV have high total chlorophyll content. Proportion of chlorophyll destroyed in drought resistant varieties is relatively less than in drought susceptible ones.
- DTSV have a lower percentage of ionizable salts as compared to drought susceptible varieties.

**Bound water*: It is a part of water held by protoplasm so strongly that it does not freeze at temperature at which free water freezes and evaporates with great difficulty even at high temperature.

- The interval between wilting and death is longer in DTSV (Singh 1964).
- Varieties with early sprouting and initial vigour are known to be drought tolerant.

8.12.5 **Measures to mitigate drought / moisture stress effects**

The loss of cane yield and quality due to moisture stress may be minimised by selecting drought tolerant sugarcane varieties and resorting to certain agronomic practices. These measures are to be applied in an integrated manner to achieve greater success.

A. Selection of drought tolerant varieties :

Selection of varieties (Table 8.48) is the most important aspect of sugarcane production under drought conditions because varieties differ widely in their response to moisture stress. A variety is said to be drought tolerant if the yield reduction due to moisture stress is less than 10 per cent (Singh, 1983).

B. Agronomic practices

i) Setts and crop hardening: Setts can be hardened by dipping them in saturated lime solution for two hours before planting. The solution is prepared by dissolving 8 kg Klin lime in 400 litres of water. It enhances germination per cent and also gives ability to crop to withstand moisture stress (Srinivasan and Naidu, 1986). Standing crop can also be hardened by withholding irrigation water for 30 days, 90 days after planting as a result of which the crop suffers less when it is required to face actual moisture stress during summer months (Raj Kumar and Kambar, 1999).

Table 8.48: Sugarcane varieties found drought tolerant in different States of India.

Varieties	Reference
Uttar Pradesh	
CoS 510, Co 312	Singh, 1964
CoS 510	Singh & Verma, 1965
CoS 510, Co 1148, BO 17, CoS 443	Singh, 1966
BO 17	Singh, 1973
BO 54, Co 1148	Singh & Singh, 1973
Co S 510, BO 17, Co 1148	Singh & Singh, 1974
BO 17,Co 1148, Co 1125, Co 1186	Singh, 1983
CoS 510,BO 54, BO 70, Co 1148,	Agrawal, 1985
Co 975, CoS 718,CoS 730, CoS 731, CoS 767, CoS 770,CoS 802	
CoS 510, BO 17, BO 54, BO 70, Co 1148, Co 975	Agrawal & Bhardwaj, 1986
CoS 767, CoS 91269, CoS 92263	Srivastava *et al.*,1997
CoS 91269,UP 39, CoS 93278	Johari *et al.*, 1998
Tamil Nadu	
CoS 510, Co 1148, Co 853, Co 997	Singh & Reddy, 1980
Co 6304, Co 740, Co 312	Naidu *et al.*, 1983

Varieties	Reference
Co 285, Co 740, Co 617, Co 997, Co 1148, Co 312, Co 349	Venkataraman *et al.*,1986
CoC 771	Devaraj,1991
Co 6304	Ramesh & Mahadevaswamy, 1999
Andhra Pradesh	
Co 997, Co 449	Prasad & Sanandachari, 1965
Co 7219, Co 6907,85 R 186 (Haritha), 83 R 23, Co A 7602, 83 A 30, CoT 8201	Rao Mukunda *et al.*, 2000
Karnataka	
Co 740, Coi 7219, Co 7125, Co M 7114, Co 8122, Co 8145, Co 86250, Co 86011, Co 86032, Co 85002	Rajkumar & Kambar, 1999

ii) Early planting: Crop of higher age by the time summer sets in and drought occurs withstand drought better than young crops. Thus, the planting time should be so adjusted that by the time severe drought starts, the cane plants are sufficiently grown up. Studies on planting dates in Tamil Nadu indicate that February planting is better than in later months (Sundara, 1984). In Andhra Pradesh, it is desirable to plant twelve months crop in January to mitigate adverse effects of drought. The best time of planting *adsali* crop is August and September for withstanding drought in next summer (Prasada Rao, 1983).

iii) Closer planting: Under drought conditions tiller production and survival are affected adversely. Drastically reduced plant population is obtained in case of moisture stress during the formative phase. Therefore, efforts should be made to have greater initial plant population by increased planting rate through closer spacing. The objective of close planting or higher seed rate is also to establish a higher stalk population to make up the growth loss of individual stalks caused by moisture stress. Close planting results in thick canopy of the crop by which evaporation losses can be minimised besides suppressing weed growth. In areas where crop is likely to be subjected to moisture stress, a row spacing of 60 cm with higher seed rate should be followed instead of 90 cm adopted for normal situations (Agrawal and Bhardwaj., 1986, Srinivasan *et al.*, 1987 and Sharma *et al.*, 1995)

iv) Deep trench planting: Before onset of drought the roots must be trained to go vertically deep, rather than spread superficially and laterally. In order to train the roots for such deeper penetration, sugarcane should be planted in trenches. Soil moisture retentivity is increased by adding high dose of FYM, compost, pressmud or similar bulky manures in the furrows before planting and stirred well. In addition, trash may be spread in the trenches to prevent surface evaporation of moisture. Trench water infiltrates more vertically than laterally. This helps the sugarcane plant to strike deep roots and draw moisture from deeper layers, where there is always more moisture. Such roots also cover more volume of soil and this helps in more absorption of soil moisture. Deep trench system of sugarcane planting produced higher cane yields at Vadapathimangalam, Tamil Nadu (Table 8.49). This could be due to better utilization of the soil moisture by root system in trenches. Considering the economics in temrs of higher yields, it may be concluded that deep trench system of planting may be followed under moisture stress conditions.

Table 8.49: Cane yields (t/ha) under different methods of planting at Vadapathimangalam, Tamil Nadu (Crop experiences severe drought during formative phase)

Planting method	Varieties			
	Co 6304	CO 62175	CoC 671	Mean
		Plant crop		
Flat beds	66.4	72.1	44.6	61.0
Ridges & furrows	68.9	79.3	55.0	67.7
Deep trenches	100.6	99.2	63.8	87.9
Mean	78.6	83.5	54.5	-
CD (0.05)	9.3			
		Ratoon		
Flat beds	50.3	72.5	30.4	51.1
Ridges & furrows	54.7	71.2	32.9	52.9
Deep trenches	110.8	115.2	63.6	96.5
Mean	71.9	86.2	42.3	-
CD (0.05	13.4			

Source: Naidu, *et al.*,1983

Table 8.50: Pattern of soil moisture availability (%) in a clay loam soil during May-June under sugarcane cropping as affected by various moisture conservation practices at Sirugamani, Tamil Nadu

Moisture conservation practices	Days after moisture stress					
	0	10	20	30	40	50
Trash mulching	100	92.6	77.5	61.5	44.6	29.5
Coir pith application	100	90.2	77.2	61.4	44.4	30.5
Deep trenching	100	91.5	76.6	63.5	47.5	31.8
Control	100	88.7	70.1	58.0	40.0	25.6

Source: Manoharan *et al.*, 1990

v) Use of soil amendments/moisture absorbants: Organic amendments (FYM, pressmud and coir waste application) improve yield of cane under moisture stress conditions by increasing moisture availability (Table 8.50) and also improving soil physical conditions due to reduction in bulk density and making soil loose and friable which enables better root spread and water and nutrient absorption.

Apart from organic amendments, use of moisture absorbants (*Jalshakti,* Stocksorb and Alcosorb) have also been found to play on important role under moisture stress conditions (Tables 8.51 to 8.53). These polymers have the unique property of absorbing water many times their weight and making it available to the plants subsequently resulting in overcoming moisture stress.

Table 8.51: Effect of soil amendments / moisture absorbants on cane yield in sandy loam soil under moisture stress and no stress conditions at Cuddalore, Tamil Nadu

Amendment/ moisture absorbant	Soil moisture		Mean
	No stress	Stress	
FYM (12.5 t/ha)	127.7	81.0	104.4
Pressmud (25 t/ha)	133.5	85.4	109.5
Jalshakti (5 kg/ha)	138.8	93.8	116.3
Coir waste (25 t/ha)	143.0	112.9	128.0
Mean	135.7	93.2	-

CD (0.05) Moisture stress (M) 1.85, Amendments (A) 1.79, MA 2.93

Source: Durai *et al.*, 1996

Table 8.52: Effect of moisture conservants on cane yield at Cuddlore, Tamil Nadu

Moisture conservant	Cane yield (t/ha)
Alcosorb (10.0 kg/ha) basal	76.6
Jalshakti (10.0kg/ha) basal	93.4
Trash mulching within a week of planting	95.5
Control	64.7
CD (0.05)	4.1

Source: Durai, 1997

Table 8.53: Effect of application of Stocksorb (S) and irrigation interval (I) on cane yield (t/ha) in medium black soil at Pune, Maharashtra.

Stocksorb	Irrigation interval (days)			Mean
(kg/ha)	10-11	20-22	30-32	
	Plant cane (Co 8014)			
0	95.6	90.2	84.6	90.1
15	99.0	96.0	93.4	96.1
20	112.3	114.4	104.8	109.2
25	112.0	109.4	103.7	108.3
Mean	103.6	100.0	95.7	-
CD(0.05)	I= 7.8, S=5.2, IS=3.6			
	Ratoon (Co 86032)			
0	78.7	72.2	65.8	72.2
15	83.0	75.8	74.9	77.9
20	86.8	82.2	78.4	82.4
25	85.2	83.2	78.0	82.5
Mean	83.3	77.8	73.5	-
CD(0.05)	I= 4.1, S= 4.7, IS 3.0			

Source: Bendigeri and Pawar, 1997

Note: Stocksorb was applied at earthing up, 95 days after planting

vi) Trash mulching: Under trash mulching evaporation is reduced due to the resistance offered by mulch covering to water vapour flow from soil surface to atmosphere and by increasing the thickness of relatively non-turbulent air zone above the soil. Due to this property, trash mulching has been tried as moisture conservation measure under drought condition in sugarcane fields by several investigators at several places. Application of trash mulch soon after emergence was found very effective in reducing water evaporation losses from soil (Table 8.54) and improving cane yields (Table 8.55). Trash mulching in ratoon crop at its initiation stage leads to conservation of soil moisture resulting in increased ratoon yields. When drought conditions prevail during grand growth period (June-October) and irrigation facilities are not available, dry and partly dry leaves on the lower portion of the cane may be removed and spread between the cane rows on the ground to prevent moisture evaporation from the soil. Trash mulching in current crop improves soil structure and water retention characters in the subsequent ratoons.

Table 8.54: Effect of trash mulching on available soil moisture (ASM) in sugarcane field during June at Sirugamani (Trichy), Tamil Nadu.

Treatment	ASM (%)					
	Days after irrigation					
	1	5	10	15	20	25
Trash mulching	100	84.0	66.0	41.0	36.7	33.0
No Trash mulching	100	73.0	49.5	36.7	31.5	29.7

Source: Parmeswaran and Ramakrishanan, 1987

Table 8.55: Yield of sugarcane as affected by trash mulching (TM) under drought conditions

Treatment	Cane yield (t/ha)	Location	Soil type	Variety	Reference
TM	125.8	Trichy,	-	CoC 671	Parameswaram &
No TM	118.5	Tamil Nadu			Ramakrishnan, 1987
CD (0.05)	1.9				
TM	82.6	Bobbili,	-	Co 6907	Ramakrishna Rao
No TM	62.0	Andhra			*et al.*, 1988
CD(0.05)	5.0	Pradesh			
TM	73.1	Melala-thur,	Clay	Co 6304	Jayabal &
No TM	65.7	Tamil	loam		Chockalingam,
CD(0.05)	5.3	Nadu			1991
TM	110.2	Siruga-mani,	Clay	Co 6304	Manoharan
No TM	104.6	Tamil	loam		*et al.*, 1990
CD(0.05)	2.4	Nadu			
TM	67.7	Dharmapuri,	Black	Co 6304	Kathiresan &
No TM	53.9	Tamil	loam		Balasubramanian,
CD(0.05)	4.6	Nadu			1991
TM	95.5	Cudda-lore,	Sandy	-	Durai, 1997
No TM	64.7	Tamil	loam		
CD(0.05)	4.1	Nadu			

vii) Spraying of potassium: Experiments have clearly proved that potassium application is very useful under early drought conditions, because potassium plays an important role in processes like transpiration, respiration, translocation of sugars and carbohydrates, energy transformation and enzyme actions (Nageswara Rao, 1987). Potassium in plant system regulates stomatal opening and closing by which it governs the transpiration losses. It maintains the turgidity of plant cells and low availability of potassium decreases moisture content in plants. Therefore, to maintain moisture content in plants at optimum level, potassium application is very necessary.

Accumulation of free proline in sugarcane plant is a valuable indication of drought resistance, which is greatly increased by adding potassium chloride (Rutherford, 1989).

Potassium content tends to be lower in plants grown on drier soils, due to decreased rate of its absorption from the soil. Therefore, foliar application of potassium chloride thrice at monthly interval during drought period increases the vigour of crop, maintenance of turgidity and survival of tillers resulting in increased cane yield. Sugar yield can further be increased if urea is added in solution of potassium chloride (Table 8.56). The solution of potassium chloride and urea should be of 2.5% each (25kg KCL+25 kg urea in 1000 litres of water/ha).

viii) Use of chemicals and anti-transpirants: Use of chemicals which induce tillering or anti-transpirants had very little commercial benefit (Srinivasan and Naidu, 1986) in sugarcane grown under drought conditions.

8.13 Waterlogging

Areas of high rainfall, due to poor drainage both surface

Table 8.56: Cane and sugar yields as affected by foliar sprayings of potassium(60, 90 and 120 DAP*) alone and with nitrogen during drought period at Coimbatore, Tamil Nadu

Spraying of	Yield (t/ha) Cane	Sugar
Sandy loam soil		
Potassium chloride	90.0	6.77
Potassium chloride +Urea	85.2	7.66
No spraying	68.8	5.28
Loam soil		
Potassium chloride	95.0	6.93
Potassium chloride +Urea	89.3	7.47
No spraying	89.7	6.60

Source: Naidu *et al.*, 1983 *DAP= Days after planting

(Note: Water solution of potassium chloride and Urea, 2.5% concentration of each)

and of the subsoil, experience waterlogging during *monsoon* period. A considerable area under sugarcane crop in several parts of India (Assam, Bihar, West Bengal, eastern Uttar Pradesh, coastal regions of Andhra Pradesh, Tamil Nadu, Kerala and Karnataka) is exposed to stagnant water for two to three months during *monsoon* season. Generally waterlogging coincides with grand growth phase to maturity of the crop.

Among the cultivated crops, cane is found to withstand the stagnation of water for a long period and accordingly the cultivation of cane in areas liable to floods and waterlogging is a common practice. So long as growing point and the top most leaves are above the level of the flood water no permanent damage is caused to the growth of the cane crop. Crop picks up the growth as soon as the flood water subsided. Although the yield and quality is reduced drastically, sugarcane is a crop that can offer at least some assured returns to the farmers under such situation.

18.13.1. Changes occurring in soil due to waterlogging

a) Depletion of oxygen: Soil-air-water relationship is an important consideration in realizing the desired growth and yield of crop. Well aerated soils have conditions which allow oxygen to enter the soil and spread through its pore spaces to meet the oxygen demand of roots and micro-organisms for respiration. Pore-space in soil if fully filled with water restricts the diffusion of oxygen into the soil. Thus, the first change that occurs due to waterlogging is depletion of soil oxygen. Some oxygen trapped in blocked pore spaces is also rapidly utilized by anaerobic organisms.

b) Inhibition of gaseous exchange: Water not only restricts the oxygen supply but also stops the escape of gases. Soil gases like carbon dioxide and methane (CH_4), accumulate and lead to plant damage.

c) Changes in the electrochemical properties of soil

i) Change in pH: The pH values of the waterlogged soils lie between 6.5 to 7.0, regardless of the initial pH values. Under such situation, the pH values of acid soils increase while those of calcareous soils decrease (because of carbon dioxide which forms carbolic acid).

ii) Changes in Eh values: Eh (Redox-potential) provides a measure of the tendency of a system to reduce or oxidize chemicals and is expressed in volts or millivolts. The well aerated soils are characterized by high positive potentials in the range of +500 to +700 millivolts. After flooding there is a sharp drop in the potential and within a few weeks negative potentials (– 300 to – 400) are observed in most of the soils depending upon type of soil organic matter, Fe and Mn content, pH, temperature and bacterial potentials.

d) Loss of nutrients: Nitrogen is lost through denitrification and leaching. Denitrification is one of the major causes of nitrogen loss from a waterlogged soil.

e) Excessive availability of Mn and Fe: In waterlogged conditions anaerobes convert the oxidized Mn and Fe compounds (Manganese oxide and Ferric oxide) into their reduced forms (Manganous oxide and Ferrous oxide) which are more soluble than their oxidized forms. Consequently, the concentration of Mn and Fe increases to toxic levels because of higher level of availability. Almost all manganese is reduced within a week of flooding.

f) Poor microbial activity: It is due to anaerobiosis, condition in which oxygen supply is lacking. The most apparent evidence of the effect of waterlogging on microbial activity is the slow decay of plant residues.

8.13.2 Effect of waterlogging on sugarcane plants

i) Morphological changes: With reduction in overall growth (Table 8.57) certain morphological changes occur in roots, stalks and leaves of sugarcane plants under waterlogged

Table 8.57: Growth characteristics of flood survived canes of different varieties at Pusa, Bihar (Crop was submerged in water to depth of 100-120cm)

Variety	Condition	Cane height (cm)	Cane girth (cm)	Inter-nodes /cane	Green leaves/ clump	Leaf area (cm^2)	Leaf length (cm)	Leaf width (cm)
Bo 11	Flooded	116.6	4.50	22.8	6.9	3305	111.5	3.9
	Non-flooded	148.2	6.05	23.2	8.4	3665	139.5	5.3
Co 453	Flooded	168.1	4.32	19.0	6.9	2508	120.8	3.2
	Non-flooded	255.0	5.21	20.4	7.7	4240	127.7	4.8
Bo 32	Flooded	179.3	4.67	20.0	8.6	2687	105.7	3.4
	Non-flooded	238.8	5.33	21.9	7.9	3755	142.7	4.9

Source: Pandey, 1964

conditions. The entire underground root system gets choked (Fig. 8.8) and adventitious roots emerges from nodes, particularly at water level (Fig. 8.9-8.11). These roots are adapted to waterlogging conditions than the original roots because they have much larger intercellular spaces. The leaves tend to become yellow with drying of lower leaves. Leaf area is

Fig. 8.8 : Blackening of underground roots due to waterlogging

Fig. 8.9 : Emergence of adventitious roots from the nodes at water level

Fig.. 8.10 : Adventitious roots (white above) and underground roots (black below) as a result of waterlogging

reduced due to reduction in number of green leaves caused by drying of lower leaves and reduced rate of leaf unfurling as well as decrease in length and width of leaves. There is a high rate of stalk mortality with considerable reduction in stalk elongation which survives. Cane height, girth and number of millable canes is decreased which results in reduced cane yield. Profuse flowering takes place in sugarcane under waterlogged conditions and crop suffers from pithiness, high fibre content and sprouting of aerial buds specially after arrowing.

ii) Physiological changes: Physiological changes

Fig. 8.11 : Presence of adventitious roots (white) indicates the level of waterlogging in field

include reduction in water and nutrient absorption and other metabolic activities of plant like transpiration, photosynthesis and respiration. The absorption of water and nutrients is influenced by the rate of root respiration and cell permeability. The energy of respiration is utilized in nutrients and water absorption. Since root respiration and cell permeability are reduced due to oxygen deficiency, water and nutrients' absorption is also slowed down. Under such situation, plants exhibit nutrient deficiency and apparent wilting symptoms on soils even well supplied with available nutrients and water.

Decline in relative growth rate and net assimilation rate have also been recorded. Chlorophyll, nitrogen, potassium and sheath moisture decreased (Fig. 8.12). Invertase activity increased leading to the hydrolysis of disaccharide into simple sugars. Such mechanism may enable sugarcane plant to

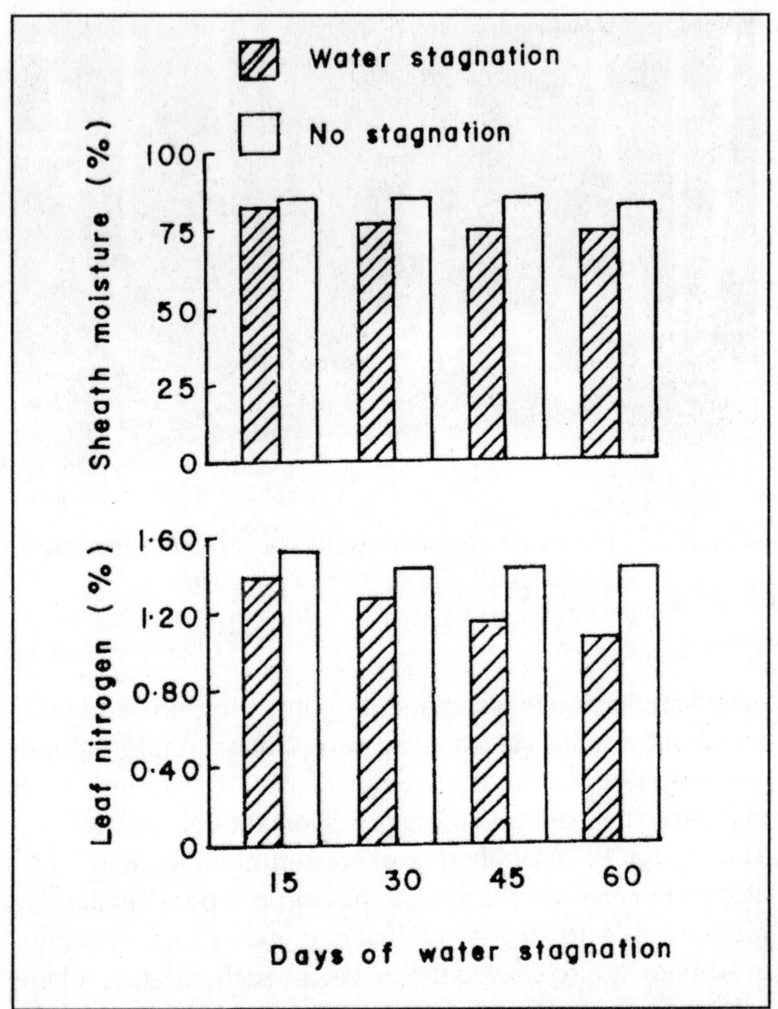

Fig. 8.12 : Sheath moisture and leaf nitrogen percentage in sugarcane as affected by different durations of waterlogging at Sirugamani, Tamil Nadu.
(5-10 cm submergence was maintained from 225-285 days of crop stage, coinciding with grand growth to early maturity phase)

Source : Manoharan *et al.*, 1990

survive under waterlogging conditions.

iii) Cane quality: There is adverse effect of prolonged flooding on sugar recovery (Table 8.58) and the difficulties created in processing. Polarisation showed a large drop compared with normal cane. The content of reducing sugars, total colloids, gums, pectin, non-protein nitrogen and ash increased but phosphate content reduced.

Table 8.58: Juice quality of different sugarcane varieties surviving flooding as compared to their healthy counterparts at Busa, Bihar (Crop was submerged in water to depth of 100-120 cm)

Variety	Condition	Bix (%)	Sucrose (%)	Glucose (%)	Purity (%)
Bo 11	Flooded	22.6	16.36	1.84	81.8
	Non-flooded	23.6	15.23	0.40	72.3
Co 453	Flooded	17.1	15.12	1.55	76.4
	Non-flooded	21.7	14.09	0.42	70.5
Bo 32	Flooded	19.9	16.25	1.69	92.1
	Non-flooded	21.8	14.63	0.36	71.1

Source: Pandey, 1964

The juice sucrose content of crop subjected to water stagnation will be higher than normal crops early in the season but the former will not attain the same peak sucrose as normal crops. Storing quality of jaggery from such crops is noted to be inferior to that of normal crops.

iv) Increased susceptibility to pests: Waterlogging enhances the susceptibility of sugarcane crop to diseases specially red rot and insect pest like white fly.

v) More lodging: Loss of vitality of roots besides adversely affecting uptake of nutrients and water also reduces the anchorage of the plant. As a result, crop becomes vulnerable to lodging and uprooting by even winds of moderate velocity.

vi) Poor ratooning ability: The ratoons raised from crop having suffered from waterlogging is usually poor due to damage caused to stubble roots.

vii) Aftereffect of waterlogging: Sugarcane maintains its juice quality so long as water remains in the field but deteriorates rapidly when water recedes. Canes of waterlogged areas also start drying at a faster rate after water receds in comparison to normally grown canes. Therefore, the flood-survived canes should be harvested as early as possible after water recession so that maximum amount of sucrose is obtained.

8.13.3 **Practices to combat waterlogging problems**

i) Drainage: Land drainage work in waterlogged tracts may be considered in two steps:

a) Construction of outlet, and

b) Field drains

Outlet drainage work is the first major step to be given consideration because it is impossible to secure good drainage in the field without provision of adequate outlet. In our country, in waterlogged areas, there is no artificial system in vogue to permit drainage of flood/water except natural outlets and slopes. However, drainage is as important as irrigation to sugarcane because the crop passes through drought period in early growth phases during summer months and waterlogging in grand growth phase during *monsoon* period. Under sugarcane growing conditions involving small and/or scattered fields, it is very difficult to plan viable project for drainage. Moreover, when inundation of the entire region takes place and the area is converted into a trough, drainage of water through gravity flow will be physically difficult.

ii) Waterlogging tolerant varieties: Sugarcane varieties differ considerably in degree of damage caused to them due to waterlogging (Tables 8.59 to 8.64). Varieties having following characteristics perform well under waterlogged conditions.

a) Produce leaf at a faster rate

Table 8.59: Performance of sugarcane varieties under waterlogged conditions at Ghaghraghat, Baharaich, Uttar Pradesh (Mean of data from 1988-89 and 1989-90)

Variety	Millable canes (000/ha)	Cane yield (t/ha)	CCS (t/ha)
CoS 8118	178	83.0	8.52
BO 91	129	74.4	7.83
CoS 767	113	67.9	7.23
Co 7918	132	64.8	6.84
CoS 8009	106	60.7	6.34
CD(0.05)	10.5	2.8	0.38

Source: Singh *et al.*,1991

Table 8.60: Performance of sugarcane varieties under waterlogged conditions (20 cm depth of water from June-Sept) at Buralikson, Assam.

Variety	Yield (t/ha)	CCS %
Co Bln 9103	46.5	14.23
Co 740	45.5	13.38
Co 8232	41.7	13.16
Co Bln 9101	41.2	14.28
Co 8340	39.7	13.79
Co Bln 9102	39.5	13.07
Co 6304	36.9	11.97
Co Bln 9104	33.1	13.65
Co 1148	32.6	12.95
Co 961	29.2	12.82

Source: Dey and Singh, 1994

Table 8.61: Performance of sugarcane varieties under waterlogged conditions (30 cm depth of water from June to mid October) at Seorohi, Uttar Pradesh (Mean of data from 1983-84 and 1984-85).

Variety	Cane yield (t/ha)	CCS (%)
BO 91	37.8	11.2
BO 89	33.4	11.1
CoS 796	26.4	10.8
CoS 767	24.1	11.5
BO 70	21.4	10.4
UP 1	18.4	10.3
CoS 802	17.8	9.4
CoS 837	15.7	10.9
CD(0.05)	9.0	-

Source: Lal and Sharma, 1988

Table 8.62: Performance of sugarcane varieties under waterlogged conditions (20-45 cm depth of water during August–Sept) at Pusa, Bihar

Variety	Cane yield (t/ha)	Sucrose % juice
BO 128	79.6	17.22
CoP 9103	68.1	17.20
BO 110	63.7	17.22
CoP 9104	59.6	17.52
CoP 9206	59.6	17.29
BO 126	58.2	17.27
CoP 9102	56.8	17.22
BO 91	54.4	17.11
CoP 9106	50.0	17.20
Co 1148	48.6	17.03
CoS 90230	46.2	17.39
CoS 90237	44.8	17.24
CoS 90217	44.3	16.92
CoP 9105	43.3	17.34
CoS 767	43.1	17.48
CD (0.05)	5.1	NS

Source: Singh & Kumar, 1978

Table 8.63: Effect of waterlogging on sugarcane yield at Thiruvalla, Kerala (waterlogging from June to October)

Variety	Yield (t/ha)		
	Unflooded	Flooded	% decrease
Co 62175	135.6	37.8	72.1
Co 449	92.8	29.4	68.3
Co 785	72.8	35.6	51.2

Source: Neelakantan *et al.*,1985

Table 8.64: Yield and quality of sugarcane varieties as affected by duration of waterlogging at Sirugamani, Tamil Nadu

Variety	Days of water stagnation*				No water stagnation
	15	30	45	60	
			Cane yield (t/ha)		
Co Si 86071	120.3	117.1	95.7	89.7	122.1
Co Si 776	118.6	114.0	92.1	85.6	120.4
CoC 671	110.3	107.8	85.4	76.1	119.8
Co 6304	108.7	105.1	86.1	76.1	116.5
CD (0.05)	4.1	2.1	1.9	1.3	3.1
			CCS %		
Co Si 86071	12.50	12.08	11.74	11.06	12.58
Co Si 776	13.42	13.12	12.60	11.82	13.96
CoC 671	13.76	13.25	12.10	12.86	14.26
Co 6304	12.06	11.62	11.53	10.27	12.63
CD (0.05)	1.03	1.22	1.31	0.94	1.23

Source: Manoharan *et al.*,1990

(*The height of submergence was artificially maintained from 5 to 10 cm for two months, from 7½ to 9½ month stage of crop, coinciding with grand growth to early maturity phase)

b) Varieties with comparatively quicker growth rate.

c) Greater ability to develop adventitious roots and porous tissues/intercellular spaces in existing roots. Several varieties develop white and turgid roots from nodes of stem, particularly at water level.

d) Slow and delayed decline in nitrate reductase activity (NRA) during water stagnation period and resumption of such activity to the normal level when water recedes.

Varieties found tolerant to waterlogging for different States are:

Uttar Pradesh: CoS 767, CoS 8118, BO 91, CoS 837

BO 54, CoS 8016, CoS 633, Co1007,

Co 1157, Co P1, Co P2,

Haryana: Co 7717, CoS 767

Bihar: BO 128, BO 17, BO 43, Co P1, Co P2, CoP 9103

Andhra Pradesh: Co 975, Co 419

Kerala: Co 775, Co 62175

Assam: Co Bln 9103, Co 740, Co 8232, Co 87025

iii) Early planting: Cane planting should be done as early as possible in fields prone to water stagnation during *monsoon* months to avoid waterlogging at juvenile phase of crop growth. The earlier planted crop attains considerable height and vigour by the time flooding occurs and will be in position to tolerate the adverse effect of flooding. In subtropical India planting of cane in autumn proved more beneficial than spring under such situation.

iv) Planting on ridges/raised beds: By planting

sugarcane on ridges/raised beds with adequately spaced subsidiary and main drains, moisture content of the soil volume in which most of the sugarcane roots are concentrated, can be lowered. As a result of this air will penetrate into the soil more easily and become available to the roots of the plant. At Lucknow, ridge/raised bed planted crop produced significantly higher cane yield than flat planting (Table 8.65).

Table 8.65: Cane yield as affected by methods of planting under waterlogging condition at Lucknow, Uttar Pradesh (15 cm standing water in the field during August and September)

Planting method	Cane yield (t/ha)		
	1982-83	83-84	Mean
Raised bed	59.8	71.2	65.5
Ridge	57.8	71.9	64.9
Flat	47.8	60.4	54.1
Flat+earthing up	52.1	66.1	59.1
CD (0.05)	2.5	2.4	-

Source: Prasad and Yadav, 1991

v) Earthing up: In case of shallow waterlogging, irrespective of method of cane planting, earthing up of cane crop before *monsoon* must be resorted to, so that excess water is drained out through furrows.

vi) Early harvesting: Waterlogged cane attain higher juice sucrose content early in the season which does not decrease as long as water is present in the field. But once the water recedes the juice quality deteriorates rapidly and the cane plants start drying up at a faster rate. Therefore, as a management measure, flood surviving canes should be harvested early in the crushing season.

vii) Spraying of cycocel: Pre-harvest spraying of cycocel (2500ppm) on waterlogged canes resulted in 1.52 units more recovery in comparison to canes which were not sprayed with cycocel in north Bihar (Sharma *et al.*, 1978).

9. Ratoon decline

Experience over the years indicates that, in general, there is a gradual decline in cane yields with an increase in number of ratoons. However, the rate of decline depends upon numerous factors like soil and climatic conditions in which the crop is being grown, variety, intensity of insect pests and diseases as well as the level of crop management. Ratoon decline is a complex problem and our study on this aspect is far from complete. However, some of the important causes of ratoon decline are mentioned below.

9.1 **Formation of toxic substances in the rhizosphere:** While identical plants are monoculturally grown in the same soil, the root exudates of living plants and/or decomposing plant residues often provide many organic and inorganic substances to the soil. These substances stimulate certain kind of microbial growth which may produce phytotoxic substances inhibiting the growth of the plants. Such phytotoxic effect due to long term monoculture of sugarcane had also been reported by many worker (Mills and Vlitos, 1965; Wu *et al.*, 1975; Shipton, 1977; Li *et al.*, 1982a, 1982b; Kao *et al.*, 1983; Kao and Hsich, 1986; Kao and Hsich, 1989). Kao and Hsich (1986), at Taiwan, found that multiple ratooning in sugarcane caused imbalance of soil fungi which led to the predominance of *Fusarium oxysporum* and *Trichoderma harzianum*, producers of phytotoxic substances. Bioassay showed that the culture filterate of F. *oxysporum* and T. *harzianum* significantly inhibited the growth of sugarcane. Settled cell volume tests and oxygen uptake experiments indicated that these two fungal metabolites significantly inhibited respiration and metabolic processes of sugarcane cells, and this retarded cell division and production.

Kao and Hsich (1986) also studied the effect of application of cane residues (roots and stubbles) on the rhizosphere fungi and the growth of sugarcane. Application of cane residues before sugarcane planting significantly affected the fungal flora of rhizosphere and the yield of cane. The populations of *Fusarium* and *Trichoderma* spp. in the residue treated soil were 2-10 fold those of non-treated soil. It greatly inhibited plant growth resulting in 8-21% decrease in cane yield. Soil inoculation tests and phytotoxicity bioassay revealed that retardation of growth of sugarcane was not attributable to the cane residues themselves, but to the rapid multiplication of F. *oxysporum* and T. *harzianum* in the residue treated soil, and production of inhibitory metabolites by these fungi.

9.2 **Depletion of soil nutrients reserves:** Growing plants of the same species continuously for a longer period in the same field put heavy demands on certain type of plant nutrients from the same soil layer. Besides depletion of soil fertility, it may also disturb the desired nutritional balance required for the proper growth of the crop plants. In studies of Mohan Rao *et al.* (1956) at Anakapalle, nitrogen content in soil (Table 9.1) at any stage of sugarcane growth was lower in succeeding crop as compared to preceding ones in multiple ratooning. It was also observed that the yield of first ratoon could be raised considerably by increased application of nitrogen but the yield of subsequent ratoon could not be pulled up to the level of the first ratoon by increasing nitrogen doses. It, thus, indicates the need for other elements in addition to nitrogen. Evidences show that organic manuring increased yield and arrested the yield decline (Yadava, 1992) in multiple ratooning which point to the importance of balanced

Table 9.1 : Total nitrogen (%) in soil during growing period of successive sugarcane crops at Anakapalle (Andhra Pradesh, India)

Crop	Month					
	March	April	June	August	October	February
Plant Crop	0.084	0.082	0.113	0.071	0.054	0.064
I Ratoon	0.069	0.078	0.110	0.069	0.052	0.059
II Ratoon	0.065	0.069	0.104	0.055	0.051	0.060

Source: Mohan Rao *et al.*, 1956

manuring in multiple ratooning of sugarcane.

9.3 **Inherent lesser physiological capacity of the succeeding ratoons for nutrient uptake :** The uptake of nitrogen, phosphorus and potash decreases with the increasing number of ratoons (Table 9.2). This shows inherent physiological incapacity of succeeding ratoons to take up nutrients. Similarly, leaf analysis which is adopted for diagnosing the nutrient status of the cane plant, also indicated that nitrogen content in leaves decreased progressively in

Table 9.2 : Uptake (kg/ha) of N, P_2O_5 and K_2O by successive sugarcane crops (Cv. Co 419) at Anakapalle (Andhra Pradesh, India)

Crop	Nutrient uptake (kg/ha)		
	N	P_2O_5	K_2O
Plant Cane	163.1	201.1	398.1
I Ratoon	126.4	108.2	342.8
II Ratoon	86.0	59.0	224.1

Source: Mohan Rao *et al.*, 1956

successive ratooning at all the stages of growth (Table 9.3).

Table 9.3 : Leaf nitrogen (%, dry basis) in successive crops of sugarcane (Cv Co 419) during different growth stages at Anakapalle (Andhra Pradesh, India)

Crop	Month after planting/starting		
	3rd	5th	8th
Plant Crop	2.53	1.86	1.51
I Ratoon	2.19	1.72	1.43
II Ratoon	2.11	1.66	1.46

Source: Mohan Rao and Narasimhan, 1954

9.4 **Upward shifting of root system:** The shoots of successive ratoon plants originate at a higher level than those of preceding ones. Consequently, root system of sugarcane becomes shallower and shallower with the increase in number of ratooning. The limited soil volume occupied by the shallow root system cannot supply enough water and nutrients for the requirement of sufficient plant growth.

9.5 **Loss of tilth :** The loose and friable soil at the beginning of the plant crop gradually becomes firm and compact in successive ratooning due to alternate wetting and drying and also compaction caused by labourers or machine during the harvesting of the crop. As a result, ratoon experience inadequate soil aeration. This condition also decreases intake of water and restricts nutrients uptake by roots.

9.6 **Higher incidence of diseases and insect pests:** The build up of diseases like smut, grassy shoot, ratoon stunting and also red rot is cumulative and with the increase in number of ratoons these diseases also increase (Srivastava and Seth, 1990). Thus, with the increasing number of ratoons the yields become progressively poorer. Insects pests associated with underground stubbles, namely, cicada, termite, root borer complex and beetle grubs may also be, in part, responsible for ratoon decline (Shrivastava *et al.* 1992). Incidence of scale insect (*Melanaspis glomerata* green) and root grubs (*Holotrichica serrata* F.) generally increases with successive ratooning (Prasada Rao, 1980).

10. Number of Ratoons

It is difficult to say categorically as to how many ratoons can be taken successfully. This largely depends upon soil and climatic conditions, intensity of insect pests and diseases and also the level of crop management. In India, at least one ratoon is taken in most parts of the country. In some areas, such as western Uttar Pradesh, Haryana and Kurnool (Andhra Pradesh) the practice of multiple ratooning is in vogue (Singh, 1981). Hunsigi (1989) attempted to analyse the favourable factors responsible for multiple ratooning.

These are :

(a) Moderate climatic conditions without extremes of heat and cold are essential to have a large number of ratoons.

(b) Soil should be fertile, deep and retentive of soil moisture but with proper drainage.

(c) Soil should possess fairly high organic matter and maintain good structure after successive ratooning for better root proliferation and extension.

(d) Agro-climatic region should be fairly free from endemic pests/diseases like red rot, smut, scales, RSD and GSD.

Due to extremes in the ambient temperature and high incidence of pests and diseases, taking more than one ratoon is not desirable in subtropical belt of India. In this belt second ratoon can be taken up only under good crop management practices. However, in tropical parts of the country, where climatic and edaphic conditions are more congenial for sugarcane growth, 2-3 ratoons are ideally possible (Hunsigi, 1982). Contrary to the expected belief, Sri D.R. Prafulla

Chandra, a progressive and innovative farmer of Hoshalli village in Shimoga district (Karnataka) has been successfully growing ratoon sugarcane of Co 449 and Co 740 varieties continuously for the past 25 years (Fig. 10.1) from a single planting which was done during 1968 on two hectares. His average cane productivity for two decades and a half vacillated between 135 to 160 t/ha which is much higher than the average national cane yield (65.5 t/ha). The method adopted by him is simple and economical . Each year, after harvest, Chandra converts dried sugarcane trash amounting to 25 tonnes per hectare into rich organic manure by mulching it in the field itself.

The success of such a large number of ratoons in his field is apparently due to the following reasons (Hunsigi, 1982):

Fig. 10.1: A view of sugarcane field of Sri Prafulla Chandra at Hoshalli village (Shimoga, Karnataka) having 25th ratoon (original crop planted in 1968)

a) High native soil fertility of the forest land that has been brought under the plough.

b) Trash mulching and its incorporation into soil in situ every year to build up soil organic matter.

c) Careful gap filling and constant efforts to maintain optimum plant density.

d) Liming as soils are acidic.

e) Judicious and timely application of organics and inorganics.

Number of ratoons in other countries : Ratooning is practiced in most of the cane growing countries of the world except Java (now Indonesia) where the system of land tenure does not permit growing of sugarcane crop continuosly for more than one season (Tandon and Misra, 1954). However, the number of ratoons (Table 10.1) taken in different countries vary to a great extent.

Table 10.1 : Number of ratoons being taken in different major sugarcane growing countries of the world

Country	Number of ratoons
Egypt, Australia, South Africa, China, Brazil, U.S.A. (Louisiana, Florida, Hawaii, some parts of Puerto Rico), Philippines	2-3
Barbados, Ethiopea, Tanzania, Jamaica, Trinidad, Guana, Argentina, Venezuela	4-5
Mauritius, Cuba, some parts of Pueto Rico	6-7

It is of interest to mention here that in east China province of Fujian, cane planted in 1727 AD on 0.3 hectare land is still being grown as ratoon. The canes are of good height (1.6 m) and other characters even after two and half centuries of continuous cultivation. This is probably the world's oldest cultivated ratoon cane (Sang, 1980)

Cited References

Adhate, Shankar Rao (1960). The best ratoon. *Indian J. Sug. Res & Dev.* **IV**(3): 137-143.

Agrawal, D.C. and Gupta, R.A. (1960). Time of harvesting of plant cane for keeping ratoon in western Uttar Pradesh. *Proc. Ann. Conv. STAI* **28**(1): 1-6.

Agrawal, M.L. (1981). A better way to manage sugarcane ratoon crops in Uttar Padesh. *Indian Sug. Crops J.* **7**(3): 1-5.

Agrawal, M.L. (1985). Research work on sugarcane grown under rainfed and deficient irrigation condition in recent years in Uttar Pradesh. *Maharashtra Sug.* **10** (4): 37-39.

Agrawal, M.L. and Bhardwaj, K.M. (1986). Rainfed culture of sugarcane II. Performance of improved cane cultivars. *Bharatiya Sug.* **11** (5) : 21-26.

Agrawal, M.L. and Bhardwaj, K.M. (1986). Rainfed culture of sugarcane I. Effect of method of planting, spacing between rows of cane and levels of manuring. *Bharatiya Sug.* **11** (4): 31-41.

Agrawal, M.P. and Lal, Nand (1984). Mechanism of amelioration of lime-induced chlorosis by pyrite and other ameliorating agents. *Ann. Rep. IISR, Lucknow, pp*.31-32.

Agrawal, M.P. and Lal, Nand (1987). Use of pyrite, pressmud, ammonium sulphate and sulphur for the amelioration of lime induced chlorosis of sugarcane in vertisols. *Intern. J. Trop. Agric.* **5**(3-4): 175-181.

Ahmad, R. and Khan, A.Q. (1991). Effect of drought on

quality of sugarcane. *Indian Sug.* **40** (11): 821-822.

Anonymous (1941). Ann. Rep. of the Sugarcane Research Work at Shahjahanpur, U.P. during 1940-41.

Anonymous (1950). " Shorter Oxford Dictionary".

Anonymous (1952). Final report on the work done in the sugarcane ratooning scheme at Kalai (Aligarh, Uttar Pradesh), India (January, 1939 to March, 1949).

Anonymous (1957). Why plough out. *S. African Sugar* J. **41**: 855-899.

Anonymous (1958). Effect of fertilizer applied to plant crop on the ratoons. *Ann. Rep. IISR, Lucknow.* p. 15.

Anonymous (1959, '60 and '61). Potassium fertilization of sugarcane ratoon. *Ann Rep. IISR, Lucknow,* 1958-59, p.39; 1959-60, p.44 and 1960-61, p.47.

Anonymous (1960). Appleton Revised Cuyas Spanish Dictionary, Appleton, New York.

Anonymous (1972). Raising a bumper ratoon crop. *Seminar on sugarcane held at SRS, Jalandhar.* Feb. 4, 1972.

Anonymous (1974). Effect of certain management practices on the performance of sugarcane ratoon. *Technical Report, AICRP on sugarcane,* 1973-74. pp 15-17.

Anonymous (1978). Drip irrigation in sugarcane. *Ann. Rep. TSRI, Taiwan* (1977-78), p.16.

Anonymous (1982). *Seminar on Ratoon Management, Trichanapally, January,* 1982.

Batta, R.K. (1997). On-form water management — The Indian experiences: Issues and strategies. *Bharatiya Sug.* **22** (5): 23-40.

Bendigeri, A.V. and Pawar, M.W. (1997). Use of superabsorbant (Stocksorb) under moisture stress in sugarcane cultivation. *Proc. Ann. Conv. STAI* **59**: 87-

95.

Bhattacharya, B.B, Verma, R.S. and Pathak, K.C. (1984). Effect of method of water and nitrogen supply on rooting pattern of sugarcane under multiple ratooning system. *Ann. Rep., IISR, Lucknow, p.26.*

Bhoj, R.L. (1962). Irrigation and nitrogen requirement of first year ratoons. *Indian J. Sug. Res. & Dev.* **6**(3): 131-137.

Bhoj, R.L. (1976). Improving sugarcane ratoons in Uttar Pradesh. *Proc. Seminar on Ratoon Management, STAI, Poona*, pp. 26-32.

Bhoj. R.L. and Mathur, B.K. (1965). Effect of seasons of planting and harvesting and seed characters on plant and ratoon crops of sugarcane. *Indian Sug. J.* **9**(4): 7-211.

Blaney, H.F. and Criddle, W.D. (1945). A method of estimating water requirements in irrigated areas from climatological data, Washington. U.S. Soil Conservation Service., 23.

Bose, P.K. and Thakur, K. (1977). Critical time of irrigation and nitrogen fertilization under water deficit conditions — A review of work done at sugarcane research station, Pusa. *Indian Sug.* **26** (12): 809-811.

Brown, J.C. (1956). Iron chlorosis. *Ann. Rev. Plant Physio.* 7: 171-190.

Chaudhary, D.C., Singh, S.N. and Singh, S.B.(1992). Studies on the effect of time and fractional doses of nitrogen application on yield, yield attributes, juice quality and economic return of sugarcane ratoon. Coop. Sug **24** (2&3): 99-102.

Chaudhary, O.S. and Kanwar, R.S. (1986). Irrigation requirements of different varieties of sugarcane. *Indian*

Sug. Crops J. **XIII** (4): 12-14.

Chauhan, R.S. (1992). Effect of gaps and gap filling materials on the yield and quality of ratoon crop of sugarcane. *Indian Sug.* **41**(10): 743-745.

Chauhan, R.S. (1993). Effect of height of plant cane harvesting, stubble shaving and earthing up on ratoon crop of sugarcane. *Bharatiya Sug.* **18**(3): 57-59.

Chauhan, R.S. and Motiwale, M.P. (1989). Gaps in ratoon in relation to dates of plant crop harvest. *Paper presented in National Seminar on early maturing varieties of cane and their management, IISR, Lucknow,* March, 25-26.

Chauhan, R.S. and Singh, R.N. (1986). Effect of initial tillage and earthing up on ratoon crop. *Ann. Rep. IISR, Lucknow*, p. 18.

Chauhan, R.S., Verma, R.S. and Pathak, K.C. (1984). Effect of cycocel application in improving stubble sprouting and yields of winter started ratoons. *Coop. Sug.* **16**(3): 135-138.

Chiranjivi Rao, K. and Asokan, S. (1978). Studies on free proline association to drought resistance in sugarcane. *Sug. J.* **40** (8): 23-24.

Cholve, S.G., Kadam, B.K., Jadhav, S.B. and Dixit, R.M. (1994). Efficacy of micro-irrigation methods *vis-à-vis* conventional method in *suru* sugarcane (CoM-7125) in Deccan canal areas in Maharashtra. *Proc. Seminar on " Water Management" held at Jalgaon, Maharashtra on March 13, 1994*, pp.41-47.

Daniel , K.V. and Sanjeevi Rao, L. (1980). Economic dose of nitrogen to sugarcane varieties under cultivation in Tamil Nadu. *Indian Sug.* **29**(10): 645-648.

De Kock, P.C. (1955). Iron nutrition of plants at high pH. *Soil*

Sci 79:175.

Deol, D.S. (1964). Studies on the nitrogen and water requirement of sugarcane in the central tract of Punjab. *Proc. Bien. Conf. Sug. Res. & Dev. Workers* **5**: 292-297.

Deshmukh, A.S. and Jadhav, S.B. (1997). Response of crop geometry under micro-irrigation for sugarcane. *Bharatiya Sug.* **22** (8): 43-48.

Devaraj, G. Sakunthala, V.A. and Chockalingam, S. (1991). Performance of sugarcane varieties under induced drought conditions. *Bharatiya Sug.* **16** (12): 45-46.

Dey, P.C. and Singh, S.N. (1994). Effect of waterlogging on growth, yield and juice quality performance of sugarcane genotypes. *Coop. Sug.* **25** (12): 547-550.

Dillewigin, C.V. (1952). Botony of Sugarcane. The Chromica Botanica Co. Book Dept. Watham, Mass, USA pp.123-268.

Dillewigin, C.V. (1953). Sugar August 53, pp. 31-36.

Dixit, R.S. and Bhardwaj, K.M. (1974). Response of sugarcane varieties to different dates of planting and harvesting. *Proc. STAI* **40**:29-34.

Durai, R. (1997). Studies on stress management in sugarcane. *Coop. Sug.* **29** (2). 102-104.

Durai, R., Chinnaswami, K.N. and Kumarswamy, K. (1996). Conservation of moisture through amendments under varying irrigation reigmes on sugarcane. *Indian. J. Agron.* **41** (3): 468-471.

Dutt, N.L.(1946). Report on the survey of sugarcane research in India.(ICSC, New Delhi).

Fritz, J. (1974). Effect of fertilizer application upon sucrose per cent cane. *Proc. ISSCT* **15**: 630-632

Gill, P.S. and Alam, Masood (1967). Effect of seed rate on plant density and yield of ratoon of pre-winter planted sugarcane. *Indian Sug.* **17**(6): 471-473.

Gosnell, J.M. (1971). Some effects of a water table level on the growth of sugarcane. *Proc. ISSCT* **14**: 841-849.

Grantz, D.A. and Meinzer, F.C. (1988). Mechanism of consequences of stomatal response to drought in field grown sugarcane. *Ann. Rep. Hawaiin Sugar Planters' Assoc.* pp.9-10.

Gupta, A.P. (1981). An important factor affecting yield and quality of sugarcane ratoon. *Maharashtra Sug.* **6**(9): 37-44.

Gupta, Rajendra and Yadav, R.L. (1993). Groundwater contribution to evapotranspiration of sugarcane during summer. *Coop. Sug.* **25** (3&4): 113-114.

Hapse, D.G. (1991). Role of micro-irrigation technique in improving agricultural productivity and environmental protection. *Human Resources Management Programme on Sugarcane Development held at SBI, Coimbatore on October 7-9, 1991, pp.59-91.*

Hapse, D.G., Deshmukh, A.S. and Shinde, P.P. (1991). Techno–economic evaluation of drip irrigation system in sugarcane crop. *Proc. National Seminar on " Modern irrigation techniques in sugarcane farming systems" held at VSI, Manjari (BK.), Pune on September 4-5, 1991.*

Hapse, D.G., Deshmukh, A.S. and Shinde, P.P. (1994). Role of micro-irrigation in improving sugarcane yield and quality. *Indian J. Sugarcane Technol.* **9** (1): 5-10.

Hunsigi, G. (1981). Ratoon cane management. *Proc. National Seminar on Ratoon Management, Lucknow.* **1**: 9-18.

Hunsigi, G. (1982). **Ratooning of sugarcane in India,**

Prabhat Printers, Mysore, India.

Hunsigi, G. (1985). Ratoon cane: Some innovations. *Maharashtra Sug.* **10**(9): 19-20, 23-26, 29-33, 36, 39-40.

Hunsigi, G. (1989). Ratooning in sugarcane. *Outlook on Agriculture* **18**(4): 175-184.

Hunsigi, G. and Parmeshwarappa, N. (1975). *Regional seminar on command area development of Malaprabha and Ghataprabha Projects, Dharwar, held on February, 20-22, 1975*

Hunsigi, G. and Srivastava, S.C. (1977). Modulation of ET values of sugarcane because of high water table. *Proc. ISSCT* **16** (2): 1550–1563.

Jafri, S.M.H. (1979). Prediction of sugarcane ratoon yield in alluvial soil by soil and tissue tests. *Fert. News* Oct. 79.

Jafri, S.M.H. (1983). Soil and plant indices and limits of sufficiency and deficiency of N, P and K for growing sugarcane. *Proc. Ann. Conv. STAI* **24**: Ag. 149-155.

Jafri, S.M.H. and Alam, Masood (1980). Response of ratoons to phosphate and potash with normal level of nitrogen application to plant and ratoon crop. *Ann. Rep. IISR, Lucknow,* p.46.

Jafri, S.M.H. and Alam, Masood (1982). Response of sugarcane plant and ratoon combination to applied nitrogen simply and in combination with phosphate and potassium. *Proc. Ann. Conv. STAI* **46:** 141-151.

Jafri, S.M.H., Srivastava, S.C. and Alam, Masood (1978). Sugarcane response to various rates of NPK fertilizers in artificially developed fertility. *Indian J. Sugarcane Technol.* **1**(1): 33-37.

Jambulingam, M., Murugesan, M., Patabi, S. and Srinivasan,

T.R. (1999). Effect of irrigation frequencies on sugarcane juice quality and yield of early and mid-late varieties. *Indian Sug.* **49**(6): 419-432.

Jayabal, V. and Chockalingam, S. (1990). Studies on drought management in sugarcane. *Coop. Sug.* **21** (8): 571-573.

Jorari, D., Srivastava, R.P., Singh, S.P. Agrawal, M.L. and Singh, G.P. (1998). Effect of moisture stress on growth, yield and juice quality of sugarcane *Coop. Sug.* **30** (4): 281-282.

Joshi, R.S., Patel, R.R., Panchal, G.N. and Mehta, H.C. (1981). Response of sugarcane ratoon crops to different levels of irrigation and nitrogen. *Proc. National Seminar on Ratoon Management, Lucknow.* **1**: 19-26.

Kadam, J.R. (1994). *Proc. Joint Conv. STAI and DSTA* : Ag 165-179.

Kadirvel, A.K. (1972). Varietal-cum-manurial trial on plant *vs.* ratoon crop at Sugarcane Research Station, Cuddalore. *Coop. Sug.* **3**(6): 277-282.

Kadu, I.K., Gaikwad, S.C. and Zende, G.K. (1988). Note on the comparative performance of different sugarcane varieties receiving irrigation at longer intervals throughout the crop growth period. *Proc. DSTA* **38** (I): 65-72.

Kailasam, C. (1995). **Sugarcane production manual, SBI, Coimbatore** p.47

Kanwar, J. S. (1976). Nitrogen, phosphorus and potassium in soil-chemistry, forms and availability. Soil Fertility – Theory and Practices, pp.156-158. ICAR Publication, New Delhi.

Kanwar, R.S. (1980). Improved cane production technology

in the Punjab State. *Punjab Agril. University,* Ludhiyana.

Kanwar, R.S. and Kaur, H. (1977). Improving sprouting of stubble crops in low temperature areas. *Proc. ISSCT.* **16**: 1325-1331.

Kanwar, R.S. and Kaur, H. (1981). Further studies on improving productivity of stubble crop in low temperature area of north India. *Proc. National Seminar on Ratoon Management, Lucknow.* **1**: 27-34.

Kanwar, R.S. and Sharma, R.K. (1981). Polythene bag transplanting technique for seed propagation and replanting of ratoons. *Proc. National Seminar on Ratoon Management, Lucknow.* **I**: 39-45.

Kanwar, R.S., Singh, Sarjit, Bains, B.S., Sharma, K.K. and Garcha, A.I.S. (1988). Yield and quality of sugarcane ratoon crop as influenced by various gap filling and nitrogen levels. *Bharatiya Sug.* **14**(2): 81-83.

Kanwar, R.S., Singh, Sarjit, Sodhi, R.S. and Toor, S.S. (1992). Efficient utilization of irrigation water in sugarcane under subtropical conditions. *Indian Sug.* **41** (11): 825-828.

Kao, M.M. and Hsich, T.S. (1986). Studies on the relationship between rhizosphere fungi and the growth of sugarcane. *Taiwan Sugar* **33**(1): 8-14.

Kao, M.M. and Hsich, T.S. (1989). Fungal metabolites and growth inhibitors of sugarcane and the mechanism of phytotoxicity *Taiwan Sugar*. **36**(2): 9-18.

Kao, M.M., Wang, T.S.C., Liu, W.C., Li, S.W. and Shive, J.J. (1983). The concept and improvement of the yield decline of monoculture sugarcane in the TSC farm during the past 30 years. *Report of Taiwan Sugar Research Institute,* **100** :25-43.

Kapoor, G.P. (1966). Suggestions for increasing the acre yield of sugarcane in U.P. *All India Symposium on Sugarcane Development, April 1966. Indian Sugar Mills Association, New Delhi, pp. 204-209.*

Kapoor, G.P. (1952). Ratooning of sugarcane in the special reference to manuring in west U.P. *Proc. Conv. Agril. Workers.* U.P.

Kapur, Jatinder and Kanwar, R. S. (1984). Effect of hot-weather and irrigations on the yield and quality of sugarcane. *Proc. Ann. Conv. STAI* **48**: 5-15.

Kapur, M.L. and Kanwar, R.S. (1991). Ratoon cane productivity as influenced by direct and residual effects of phosphorus and potassium fertilization. *Proc. Ann. Conv. STAI* **53:** Ag 39-46.

Kathisesan, G. and Balasubramanian, N. (1991). Management of drought in sugarcane during early growth phase. *Indian Sug.* **41** (5): 319-324.

Khandagave, R.B. and Kagi, S.S. (1998). Water management in sugarcane. *Bharatiya Sug.* **23** (5): 43-44.

Khanna, K.L. and Raheja, P.C. (1948). Respiration studies in sugarcane under partial submergence. *Indian J. Agric. Sci.* **18**: 9-18.

Kumar, Shiv and Srivastava, S.N.L. (1991). Response of sugarcane to alternate furrow irrigation and trash mulch. *Indian J. Sugarcane Technol.* **6** (1&2): 69-73.

Kumar, Siva and Srivastava, N.S.L. (1993). Response of sugarcane genotypes to planting dates and their effect on succeeding ratoon cane. *Indian Sug.* **43**(2): 101-104.

Kumaraswamy, K. (1989). Sugarcane soil management most important to get better yield. *Kisan World* **12**: 20-22.

Lakshmikantham, M. (1965). Half century of agricultural

research at Anakapalle. *Agric. Information Service, Deptt. of Agric., AP.*

Lakshmikantham, M. (1973). **Technology of sugarcane growing**. Andhra Pradesh Agril. Univ., Hyderabad.

Lakshmikantham, M., Gopalam, B. and Prasad Rao, N.N.V.V. (1971). Drought resistance in sugarcane — A new concept. *Proc. Ann. Conv.* DSTA. 223-225.

Lakshmikantham, M., Rao, G. Narasimha, Narasimham, R.L., Rao, P. Hanumantha, Subbarao, B. and Krishna, T. Sivarama (1962). Studies on the influence of irrigation and nitrogenous fertilization on tissue composition as related to cane and sugar yields. *Proc. ISSCT* **11**: 238-247.

Lal, Kanhaiya and Sharma, Subhash (1988). Studies on morphological characters associated with waterlogging resistance in sugarcane. *Proc. Ann. Conv. STAI* **51**: 75-83.

Leather, J.W. (1910). Water requirement of crops in India. *Mem. Dept. Agric.*

Lewis, C.T. and Short, C. (1958). " A Latin Dictionary" Oxford Univ. Press (Clarendon), London and New York.

Li, S.W., Kao, M.M., Shive, J.J., Liu, W.C., Wang, T.S.C. (1982 a). Sugarcane growth in relation to soil microbial balance. *Soil and Fertilizers in Taiwan*, pp. 31-36.

Li, S.W., Kao, M.M., Shive, J.J., Liu, W.C., Wang, T.S.C. (1982 b). Pre-submergence in relation to soil microbial balance and sugarcane production. *Report of Taiwan Sugar Research Institute,* **98**: 25-35.

Lindsay, W.L. and Narvell, W.L. (1978). Development of a DTPA soil test for Zn, Fe, Mn and Cu. *Soil Sci. Soc., Am.* J. **42** :421-428.

Magadum, V.M. and Divekar, M.V. (1962). Varietal resistance to drought conditions. *Proc. Ann. Conv. DSTA*. 64-68.

Malik, J.P.S., Singh, R.K., Bhardwaj, K.M. and Agrawal, M.L. (1996). Effect of irrigation methods and schedules on growth of sugarcane alongwith production of cane and sugar in western Uttar Pradesh. *Bharatiya Sug.* **21** (6); 29-35.

Mane, V.S. and Salunkhe, C.D. (1978). Studies on the sugarcane ratoons. *Indian Sug.* **28**(6): 441-446.

Manoharan, M.L., Duraisamy, K. and Krishnamoorthy, S.V., Vijayaraghavan, H. and Muthukrishnan, K. (1990). Perfomance of sugarcane varieties in waterlogged conditions. *Bharatiya Sug.* **15** (11): 39-47.

Manoharan, M.L., Duraisamy, K. and Vijayaraghavan, H. (1990). Effect of management practices to improve cane yield and quality under moisture stress conditions. *Bharatiya Sug.* **15** (10): 19-31.

Mathur, B.K. and Singh, A. (1968). Role of cultural methods in the cultivation of the ratoon crop under deficient irrigations. *Indian Sug.* **17**(11): 815-822.

Mathur, B.K. and Tripathi, K.B. (1975). Boosting sugarcane quality through nitrogen application in relation to irrigation. *Indian Sug.* **25** (5): 389-339.

Mathur, B.K., Bhadauria, V.S. and Singh, A. (1968). Study on effects of different doses of nitrogen and systems of planting on plant cane and its subsequent effect on ratoon crop. *Indian Sug.* **18**(8): 611-619.

Mathur, P. S. and Behari, Chatur (1980). Growing sugarcane in different states of India . Directorate of Sugarcane Development, Ministry of Agric., Govt. of India.

Mathur, P.S. (1986). Practical approach to improving

sugarcane ratoon management under cultivators conditions. *Coop. Sug.* **17**(9): 695-699.

Mathur, P.S. (1987). Agronomical practices for better ratoon production. *Platinum Jubilee Souvenir, SBI Coimbatore, India.* **I**: 135-139.

Mathur, P.S. and Behari, C. (1980). Growing sugarcane in different states in India. Directorate of Sugarcane Development, Govt. of India, Ghaziabad.

Mathur, R.N. and Haider, I.U. (1940). A summary of five years of physiological research on sugarcane, Sugarcane Research Station, Shahjahanpur. *Proc. Ann. Conv.* STAT **9**( ):

Michael Raj, S., Srinivasan, T.R. and Venugopal, K. (1982). Planting systems and ratoon crop yields in problem areas. *Seminar on ratoon management, Trichnapally, January.*

Mills, J.T. and Vlitos, A.J. (1965). The rhizosphere of sugarcane: preliminary isolation of fungi from an uncultivated, heavy clay soil and from similar soil in which sugarcane is being cultivated. *Proc. ISSCT* **12**: 125-136.

Misra, A. and Mathur, P.S. (1983). Ratooning sugarcane in India – Retrospect and prospects. *India -Sug. Crops J.* **9**(3): 1-14.

Misra, A. and Mathur, P.S. (1990). Effect of gap filling materials on the yield and quality of ratoon crop in early varieties of sugarcane. *Indian J. Agron.* **35**(3): 258-261.

Misra, G.N. (1953). Comparative studies in the development and growth of sugarcane ratoon crop in relation to late season tillers and after harvest activated bud tillers. *Proc. Ann. Conv. STAI* **22**. (II): Ag. 96-101.

Misra, G.N. (1966). Early application of nitrogen as a means to mitigate its adverse effect on the juice quality of sugarcane. *India Sug.* **16**(6): 469-472.

Mohan Rao, N.V. and Narasimham, R.L. (1954). Biochemical studies on sugarcane ratoons. *Proc. Bien. Conf. Sug. Res. & Dev. Workers* **2**(II): 388-395.

Mohan Rao, N.V., Rama Rao, G. and Jagannadha Rao, E.(1956). Studies on sugarcane ratoon. *Proc. ISSCT* **9**(I): 233-254.

Motiwale, M.P. and Singh, A.B. (1977). Effect of trash mulch in economy of water and fertilizer use in sugarcane. *Ann. Rep. Indian Institute of Sugarcane Research, Lucknow, pp. 23-24.*

Mukerji, B.K. (1952). Some important findings of the sugarcane ratooning scheme at Kalai (Aligarh, Uttar Pradesh.) *Indian J. Agric. Sci.* **22**(3): 267-270.

Nageswara Rao, P. (1987). **Potash on sugarcane uptake and utilization**. Indian Potash Publication 118.

Naidu, K.M. and Bhagyalakshmi, K.V. (1973). Relative turgidity and stomatal movement of sugarcane varieties in relation to their drought resistance. *Indian J. Agric. Sci.* **43** (11): 1016-1018.

Naidu, K.M., Srinivasan, T.R. and Raj, Michael (1983). Varietal behaviour and management practices for improvement of sugarcane yield and quality under moisture stress conditions *SISSTA Sug. J.* **9** (2): 47-52.

Naidu, R.V. and Lakshmikantham, M. (1946). Studies in tillering and arrowing in Co 421 at Anakapalle. *Indian J. Agric. Sci.* **16**(2): 139-151.

Nath, A. and Mathur, B.K. (1964). Time of manuring ratoon under deficient irrigation conditions. *Indian Sug.*

14(12): 803-804.

Neelakantan Potty, N., Sukumaran Nair, S. and Verghese, S.S. (1985). Effect of inundation on growth and yield attributes of some sugarcane cultivars. *Indian Sug.* **34** (12): 815-816.

Pandey, V. (1964). Growth and biochemical studies of flood survived canes and varietal resistance to flood injury. *Proc. Bien. Conf. Sug. Res. & Dev. Workers* **5**: 340-344.

Pandian, B.J. and Rajsekaran, S. (1991). Studies on ratoon management in sugarcane. *Coop. Sug.* **23**(3): 185-188.

Pandian, B.J., Muthukrishnan , P. and Rajsekaran, S. (1992). Efficacy of different irrigation methods and regimes in sugarcane. *Indian Sug.* **42** (4): 215-220.

Paneerselvan, R., Jayabal, V., Sankaran, N., Tamilselvan, N., and Enayathullah, S. (1992). Effect of management practices and N levels on sheath moisture per cent at different stages of sugarcane ratoon. *Bharatiya Sug.* **17**(4): 47-49.

Panje, R.R. (1964). Research Results. *Ann. Rep. IISR, Lucknow, p.3.*

Panje, R.R. (1965). How to improve the sugar content of cane. *Tech. Bull. No.1, Indian Institute of Sugarcane Research, Lucknow, p.19.*

Panje, R.R. (1967). Improvement of sugarcane in north India. *Indian J. Science and Industry* **1**(1): 48-54.

Panje, R.R. (1967). Pospects of sugarcane improvement in north India. Farmer and Parliament **2**(9): 9-12.

Parameswaran, P. and Ramakrishnan, M.S.(1987). Measures to alleviate drought conditions in sugarcane due to canal closure in wet lands: Efficacy of mulches. *Indian J.*

Sugarcane Technol. **4** (2): 113-118.

Parikh, M.M., Shrivastava, P.K., Savani, N.G. and Raman, S. (1992). Response of sugarcane crop to drip method of irrigation. *Coop Sug.* **23** (10): 673-677.

Parthasarathy, S.V. (1949). *Current Science, pp* 172-173.

Patel, H.S, Shah, T.M., Arvadia, M.K. and Patel, M.P. (1989). Effect of gap filling and nitrogen levels on yield and quality of ratoon raised on sugarcane varieties. *Bharatiya Sug.* **14**(3): 33-37.

Penman, H.L. (1948). Natural evaporation from open water, bare soil and grass. *Proc. Roy. Soc. London , Serv.* A. 193: 120-145.

Plucknett, D.L., Evenson, J.P. and Sanford, W.G. (1970). Ratoon cropping. *Adv. Agron.* **22**:285-330.

Prasad, A. and Parashar K.S. (1986). Effect of moisture regimes, methods of planting and systems of cropping on yield attributes, yield and economics of spring planted sugarcane. *Indian J. Agron.* **31** (1): 45-50.

Prasad, A. and Parashar, K.S. (1989). Effect of split application of N alone and based P on growth, yield and juice quality of first ratoon. *Indian Sug.* **38**(11): 891-893.

Prasad, B., Prasad, C.R., Singh, K.D.N., Misra, G.K. and Verma, S.N.P. (1984). Fertilizer requirement for different yield targets of ratoon sugarcane based on soil test values in calcareous soils of Bihar, *Maharashtra. Sug.* **9**(7): 29-32.

Prasad, S. R., Alam, Masood and Sharma, R.K. (1981). Water and nitrogen requirments of sugarcane ratoon. *Ann. Rep. IISR. Lucknow. p.26.*

Prasad, S.R. and Yadav, R.L. (1991). Effect of planting method on cane yield under waterlogged conditions. *Indian J.*

Sugarcane Technol. **6** (1&2): 112-113.

Prasad, S.R., Alam, M. and Singh, R.N. (1987). Role of skip-furrow irrigation in increasing water use efficiency in sugarcane under staggered-row planting. *Indian Sug.* **36** (11): 569-573.

Prasad, S.R., Srivastava, N.S.L. and Alam, Masood (1980). Irrigation efficiency in sugarcane *Indian J. Agric. Sci.* **50** (3): 252-260.

Prasad, S.R., Yadav, R.L. and Singh, K. (1988). Efficient utilization of irrigation water in skip–furrow method through trash mulch in sugarcane. *Indian J. Agron.* **33** (2): 125-130.

Prasada Rao, (1983). Package of practices for sugarcane subjected to drought in Andhra Pradesh. *SISSTA Sug. J.* **9** (2): 43-46.

Prasada Rao, K.K. (1980). Ratoon cane management. *Proc. Ann. Conv. STAI* **44**: Ag 159-172.

Prasada Rao, K.K. (1981). On ratooning sugarcane. *Proc. National Seminar on Ratoon Management,* Lucknow. I: 35.

Prasada Rao, K.K. (1983). Package of practices for sugarcane subjected to drought in Andhra Pradesh. *SISSTA Sug. J.* **9** (2): 43-46.

Prasada Rao, K.K. and Sanandachary, A. (1965). A new method for determining drought tolerance in sugarcane varieties. *The Indian Sugarcane J.* **10** (1): 22-23.

Prasada Rao, K.K., Narasimharaju, P.V.L., Seetharamasharma, S. and Ramalingaswami, K. (1979). Studies in juice quality as influenced by leaf nitrogen, sheath moisture and minimum temperatures prevailing during the three months' period prior to harvest. *Indian J. Sugarcane Technol.* **2**(1): 20-24.

Prasada Rao, K.K., Ramalingaswamy, K., Ramakrishna Rao, N. and Veerabhadra Rao, K. (1987a). Studies on macro and micro nurient composition of soils and plant and ratoon cane crops raised on the same fields in Chelluru, Anakapalle and Vuyyuru Factory zone of A.P. *Proc. DSTA* **37**:A-45-58.

Prasada Rao, K.K., Ramalingaswamy, K., Ramakrishna Rao, S. and Bapu, N. (1987). Studies on macro and micro nutrient composition of certain soils and plant and ratoon cane crops raised on the same fields in some factory areas of north coastal region of Andhra Pradesh. *Coop. Sug.* **18**(11): 743-749.

Raheja, P.C. (1946). Growth studies on *Saccharum officinarum* I. Varietal Series. *Indian J. Agric. Sci.* **16**(11): 113-138.

Raheja, P.C. (1948). Growth studies on *Saccharum officinarum* II. Irrigation series. *Indian J. Agric Sci.* 18: 95-111.

Raheja, P.C. (1956). Physiological researches in relation to agronomy of sugarcane in India. *Proc.* ISSCT **9**(1): 400-418.

Raheja, P.C. and Misra, G.N. (1954). Growth of ratoon crop as affected by spacings. *Proc. Bien. Conf. Res. & Dev. Workers* **2**(II): 404-414.

Rai, R.K., Shrivastava, A.K., Ghosh, A.K. and Shukla, S.P. (1989). Nitrate reductase activity of vivo in plant and ratoon crops of sugarcane. *Indian J. Pl. Physiol.* **32**(4): 320-324.

Raj Kumar, M. and Khuntia, S.N. (1998). Irrigation and cane quality. *Proc. Ann. Conv. STAI* **60**: 6-9.

Rajkumar, A.S. and Kambar, N.S. (1999). Drought management in sugarcane. *Bharatiya.Sug.* **24** (9): 31-33.

Ramakrishna Rao, P. (1960). Some problems connected with sugarcane ratoon cultivation. *Indian Sug.* **19**(6): 403-406.

Ramakrishna Rao, S., Raju, J.S.N., Veerabhadra Rao, K., Padmanabham, M., Sadanandachari, A. and Ramapandu, S. (1988). Effect of trash mulching and nitrogen and potash nutrition on rainfed sugarcane. *Proc. DSTA* **38**: 55-64.

Ramalingaswamy, K., Ramakrishna Rao, S., Bapu, N. and Veerabhadra Rao, K. (1986). Studies on the effect of quality of irrigation water on the uptake of nutrients, sugarcane yield and quality. *Maharashtra Sug.* **11** (4) 47-54.

Ramesh, P. and Mahadevaswamy, M. (1999). Water requirement and water use efficiency of sugarcane varieties as influenced by drought during formative phase. *Indian Sug.* 49(6): 405-409.

Rao, J.T. (1981). Sugarcane ratoon management–Problems and solutions. *Proc. National Seminar on Ratoon Management, Lucknow* **I**: 1-8.

Rao, Mukunda, Rao, K.P.S, Babu, P.S., Rao, C.P. and Shrinivas, D. (2000). Management of sugarcane under soil moisture stress / drought. *Bharatiya Sug.* **25** (4): 43-46

Rathi, K.S. and Singh, R.A. (1983). Effect of potassium application on juice and net profit of cane ratoon in light soils of central U.P. *Coop Sug.* **14**(10): 517-519.

Rathi, K.S. and Singh. R.A. (1982). Effect of nitrogen rates on stalk yield of cane ratoon in sandy loam soil of central

U.P. *Coop. Sug.* **13**(7): 317-318.

Rattan, P. and Singh, Harbans (1954). Effects of some agronomic factors on the yields and quality of ratoon crop of sugarcane in Punjab. *Proc. Bien. Conf. Sug. Res. & Dev. Workers.* **2**(II): 440-450.

Ray, S.B., Gautam, O.P. and Dastane, N.G. (1964). Water requirements of crops on soil moisture basis. *Proc.* 5th *N.E.S.A.* Irrig. Semi. New Delhi.

Reddy, M. Ramachandra and Hapse, D.G. (1986). Irrigation scheduling for sugarcane crop — Maharashtra conditions. *Bharatiya Sug.* **11** (8): 63-71.

Reddy, M. Ramachandra, Reddy, S. Chandrasekhar, Venkatachari, A. and Kulkarny H.L. (1978). Studies on the effect of nitrogen and moisture regimes on yield and quality of sugarcane. *Proc. Ann. Conv. STAI* **42:** 99-122.

Rege, R.D. and Patwardhan, N.B.L. (1945). Sugarcane varietal trial in the Deccan Canal Tract II. Performance of varieties in different planting and ratooning. *Indian J. Agric. Sci.* **15**(1): 55-62.

Rutherford, R.S. (1989). The Assessment of proline accumulation as a mechanism of drought resistance in sugarcane. *Proc. Ann. Congr. S. African Sugar Tech. Assoc.* **63**:136-141 (Sugarcane, March-April, 1990, p. 28)

Saini, S.K., Singh, P.P. and Kumar, K. (1978). Studies on the soil moisture regimes in sugarcane under shallow water table conditions of north India. *Proc. Ann. Conv. STAI* **42:**123-133.

Sang, T. (1980). The world's oldest cultivated ratoon cane. *International Sug. J.* **82**: 146.

Seetharamaiah, P., Krishnamurthy, K. and Suryanarayana, J.V.V. (1952). Manuring of ratoons in sugarcane. *Madras Agric. J.* **39**(3): 218-225.

Sethiya, H.L. and Dandsey, J.P.S. (1992). Ratooning in sugarcane. 1. Morphological study of stubbles. *Indian Sug.* **41**(10): 751-754.

Shanmugasundaram, S., Misra, A., Venugopal, K. and Ethirajan, A.S. (1981). Effect of varying plant density on performance of plant and ratoon of different sugarcane varieties. *Indian Sug.* **31**(3): 171-174.

Sharma, H.K., Kapur, A., Mann, A.P.S. and Kanwar, R.S. (1985). Effect of growth regulator on ratoon cane sprouting and yield. *Sugar Cane*. **2**: 15-16.

Sharma, H.K., Singh, O.and Kanwar, R.S. (1987). Effects of nitrogen in combination with growth regulator on ratoon of low temperature harvested sugarcane. *Proc. Ann. Conv. STAI* **50**: Ag. 81-86.

Sharma, R.K. and Verma, H.D. (1996). Effect of irrigation scheduling and other agronomic manipulations on yield and water economy in sugarcane. *Indian J. Agron.* **41** (1): 122-126.

Sharma, R.K., Verma, R.D. and Rathore, O.P. (1995). Effect of irrigation and row spacing on yield and quality of sugarcane. *Indian J. Sugarcane* Technol.**10** (2): 150-153.

Sharma, S.C., Rao, G.S.C. and Sharma, S.S. (1978). Influence of cycocel on waterlogged and lodged sugarcane — a comparative study of laboratory and big mill trials. *Indian Sug.* **28** (2): 87-92.

Sharma, S.N., Singh, S., Sharma, R.P., Parashar, K.S. and Prasad, Rajendra (1981). *Teir Chrift fur Acker undllanzenban, West Germany,*

Shinde, P.P., Deshmukh, A.S., Khalkar, S.V., Gunjal, B.V. and Hapse, D.G. (1992). Effect of different levels of nitrogen through drip on yield and quality of sugarcane under surface drip. *Proc. Int. Agril. Engg. Conf. held at Bangkok, Thailand on Dec. 7-10.*

Shinde, S.H., Dahiwalkar, S.D. and Berad, S.M. (2000). Response of sugarcane to planting technique and fertigation through drip. *Bharatiya Sug.* **25** (3): 19-22.

Shipton, P.J. (1977). Monoculture and soil borne plant pathogens. *Ann. Rev. Phytopathol*. **15**: 387-407.

Shrikhande, J.G., Bhoj, R.L. and Singh, S.N. (1958). Quoted by R.L. Bhoj. Improvement in the practices for ratooning in sugarcane - A review. *Indian Sug.* 1971, **20** (12):849.

Shrivastava, A. K., Singh, Kishan, Ghosh, A. K. and Saxena, Y. R. (1988). Leaf initiation as influenced by growth and development in plant and ratoon crop of sugarcane. *Indian J. Sugarcane Technol*. **5**(1): 5-10.

Shrivastava, P.K., Parikh, M.M., Savani, N.G. and Raman, S. (1993). Response of sugarcane crop to sprinkler irrigation. *Indian Sug.* **42** (10): 761-764.

Siag, R.K., Verma, B.L. And Deo, R. (1989). Effect of different levels of irrigation and nitrogen fertilizer on the yield and quality of plant and ratoon cane crops in north west Rajasthan. Proc. Ann. Conv. STAI **52:** Ag 39-46.

Singh, A., Lal, N. (1963). Effect of bulk density of soil on growth of sugarcane. *Ann. Rep. IISR, Lucknow* (1962-63), pp. 38-39.

Singh, Anil Pratap (1997). Characters associated with drought tolerance in sugarcane – A review. *Indian Sug.* **47**(4): 283-286..

Singh, G., Singh, U.P. and Yadav, R.A. (1991). Effect of variety and spacing on yield and quality of sugarcane in flood-prone area of eastern Uttar Pradesh. *Indian J. Agron.* **36** (4): 563-566.

Singh, G.B. and Sharma, R.K. (1973). Studies on level and time of nitrogen application in sugarcane ratoon. *Ann. Rep. IISR, Lucknow*, pp. 15-16.

Singh, G.B., Srivastava, N.S.L. and Sharma, R.K. (1975). Studies on ratoon management. *Ann. Rep. IISR, Lucknow*. pp. 27-28.

Singh, J.R.P. and Kumar, R. (1998). Evaluation of sugarcane varieties under waterlogged condition. *Proc. Ann. Conv. STAI* **60**: 3-5.

Singh, K. (1981). Management of sugarcane ratoon in India. *Proc. National Seminar on Ratoon Management, Lucknow.* **II**: 186-192.

Singh, K., Bhatia, N.L. and Behl, K.L. (1977). Manuring in sugarcane under Haryana conditions. *Indian Sug. Crops J.* **4**(3): 71-72.

Singh, M. and Agrawal, M.P. (1981). Nutrient supplying potential of stubble roots of sugarcane ratoon. *Proc. National Seminar on Ratoon Management, Lucknow*. **II**: 144-147.

Singh, N., Kanwar, R.S. and Singh, O. (1988). Effect of ethrel and emison on ratoon of cane crop harvested during low temperature period. *Coop. Sug.* **19**(6): 391-392.

Singh, N.B. and Ali, S.A. (1981). Drought tolerance in sugarcane cultivars. *Proc. Ann. Conv. DSTA*. A15-24.

Singh, O. (1990). Effect of auxin on cane ratoon of chilling harvested crop. *Bharatiya Sug.* **15**(5): 47-49.

Singh, O. and Kanwar, R.S. (1988b). Combined effect of

ethrel and lindane on low temperature ratooning of sugarcane. *Coop. Sug*. **19**(6): 381-385.

Singh, O. and Kanwar, R.S. (1988a). Ratoon of cold harvested sugarcane as influenced by exogenous application of ethephon. *Indian J. Sugarcane Technol*. **5**(1): 69-71.

Singh, O. and Kanwar, R.S. (1991). Hormonal efficiency to trigger activity of subterranean buds of sugarcane ratoon under low temperature conditions. *Proc. DSTA* (I): A1-6.

Singh, Onkar and Singh, O.S. (1974). Significance of soil moisture during simulated freezing for sugarcane survival. *Indian Sug.* **24** (5): 377-382.

Singh, P.P. and Singh, Govinda (1971). Soil moisture regime and levels of nitrogen: Effects on the yield and quality of sugarcane. *Proc. ISSCT* **14**: 853-858.

Singh, P.P., Saini, S.K., Kumar, Kamlesh and Negi, P.S. (1986). Studies on plant population and irrigation requirements of sugarcane under *tarai* conditions of sugarcane. *Indian J. Sugarcane Technol.* **3** (1): 20-23.

Singh, R.G. and Singh, S. (1974). Studies on drought tolerance of sugarcane varieties. *Proc. Ann. Conv. STAI* : **40** (I) : 73-80.

Singh, R.G., Bhardwaj, M. and Singh, H.N. (1981). Ratoon management in U.P.: A status paper. *Proc. National Seminar on Ratoon Management, Lucknow.* **III**: 223-239.

Singh, S. and Kanwar, R.S. (1965). Latest results of sugarcane research in Punjab and future lines of work. *Indian Sug.* **15**(5): 371-373.

Singh, S. and Reddy, Sanjiva M. (1980). Growth, yield and juice quality performance of sugarcane varieties under different soil moisture regimes in relation to drought resistance. *Proc. ISSCT* **XVI** (1): 541-555.

Singh, S. and Reddy, Sanjiva M. (1980). Growth, yield and juice quality performance of sugarcane varieties under different soil moisture regimes in relation to drought resistance. *Proc. IISSCT* **XVII** (1): 541-555.

Singh, S.N., Singh, S.B. and Singh, N.P. (1994). Effect of moisture regimes and methods of irrigation on growth, yield, quality and water economy in sugarcane crop . *Proc. All India Seminar on Water Management held at Jalgaon, Maharashtra on March 13, 1994, pp 7-12.*

Singh, Sudama (1964). Studies on the drought endurance capacity of cane varieties — The time lag between wilting and death under waterlogged condition. *Indian Sug.* **13** (12): 717-720.

Singh, Sudama (1967). Growth of cane in relation to soil moisture stress during elongation phase. *Indian Sug.* **17** (3): 283-286.

Singh, Sudama and Verma, H.P. (1965). Studies on drought endurance capacity of cane varieties – Relation of transpiration to the size and number of stomata. *The Indian Sugarcane J.* **9** (4): 252-254.

Singh, Sudama, (1966). Drought in sugarcane. *Indian Sug.* **16**(8):609

Singh, Sudama, Srivastava, K.K. and Misra, G.P. (1976). Critical irrigation period in sugarcane crop under limited irrigation facilities. *Indian Sug.* **26** (3): 141-143.

Singh, U.S. (1973). Studies on water relations of sugarcane — Relative efficiency of some outstanding canes to grow under normal and deficient water supply and to endure drought and soil desiccation. *Sugar News* **5** (4):

10-12.

Singh, U.S. (1983). Drought resistance in sugarcane I. Influence of drought on plant productivity. *Indian Sug.* **32**(12): 947-951.

Singh, U.S. and Negi, V.S. (1969). Performance of some improved canes under different levels of irrigation and their relative ability in enduring the stress of drought conditions. *Indian Sug.* **18** (11): 825-832.

Singh, U.S. and Singh, Malipal (1973). Response of sugarcane varieties to withholding of water for different duration in the pre-*monsoon* period. *Sugar News* **5** (1): 12-14.

Sivanappan, R.K. (1983). Drip irrigation for sugarcane. *SISSTA Sugar J.* **9** (2): 27-32.

Srinivasan, T.R. and Mariakulandai, A. (1969). Influence of N and moisture regime on crop indices and yield of cane. *Indian Sug.* **19** (6): 465-472.

Srinivasan, T.R., and Naidu, K.M.(1986). Maximizing sugarcane production under moisture stress conditions during formative phase. *Proc. Ann. Conv. DSTA* **36**: A 421-432.

Srinivasan, T.R., Naidu, K.M. and Kathiresan, M.K. (1987). Effect of certain management practices on yield improvement of sugarcane grown under moisture stress. *Proc. Ann. Conv.* DSTA **37**: 111-222.

Srivastava, A.K. (1983). The assessment of critical limits of compaction in sugarcane soils in relation to growth and yield of sugarcane. *Ann. Rep. IISR, Lucknow, pp. 43-44.*

Srivastava, Arun K. (1981). Soil compaction – its relevance in ratoon management. *Proc. National Seminar on Ratoon Management, Lucknow.* **II**: 193-200.

Srivastava, R.P., Tewari, A.N., Johari, D. and Singh, S.P. (1997). Effect of water stress on stomatal opening potential in sugarcane. *Indian J. Sugarcane Technol.* **12** (1): 52-58.

Srivastava, S.C. and Seth, S.R. (1990). Sugarcane agronomy and cropping. **Recent advances in sugarcane**, M/S the K.C.P. Limited, Vuyyuru, A.P., India, pp. 207-247.

Srivastava, S.N.L., and Kumar, S. (1992). Weed management in sugarcane ratoon, *Indian J. Sugarcane Technol.* **7**(1): 45-47.

Subramanian, K.S., Nandagopal, A. and Selvakumari, G. (1989). Irrigation need of sugarcane at critical stages. *SISSTA Sug. J.* **3**: 111-113.

Subramanian, N. (1982). Ratoon practice in Tamil Nadu. *Proc. Seminar on Ratoon Management, Trichanapally, January, 1982.*

Sundara (1996). Sugarcane varieties and management practices for saline water irrigated areas. *Bharatiya Sug.* **21** (5): 35-39.

Sundara, B. (1984). Studies on varieties and their age at harvest under restricted irrigation in semi-arid tract. *Proc. DSTA* **34**: A 138-145.

Sundara, B. and Reddy, B. Syamasundar (1994). Studies on sugarcane varietal performance and planting systems under saline water irrigation. *Coop. Sug.* **25** (11): 457-461.

Tandon, R.K. (1955). Serving Uttar Pradesh through Research in sugarcane.

Tandon, R.K. and Misra, G.N. (1954). On the behaviour of sugarcane ratoon in the particular reference to north Indian conditions. *Proc. Bien. Conf. USG. Res. & Dev.*

Workers. **2**(II): 380-388.

Tandon, R.K. and Misra, G.N. (1954). On the behaviour of sugarcane ratoon with particular reference to north Indian conditions. *Proc. Bien. Conf. Sug. Res.& Dev. Workers.* **2**(I): 25-26.

Tandon, R.K. and Misra, G.N. (1956). Time of application of nitrogenous fertilizers to sugarcane. *Proc. ISSCT* **9**(I): 139-150.

Tandon, R.K., Kapoor, G.P. and Misra, G.N. (1955). Ratooning of sugarcane as influenced by the time of harvesting plant cane. *Indian J. Agric. Sci.* **25**(1): 31-40.

Tandon, R.K., Singh, M. and Dixit, A.S. (1957). Some cultural and manurial treatments for raising a good ratoon of sugarcane crop in the eastern U.P. *Proc. Bien. Conf. Res & Dev. Workers* **3**(II): 130-135.

Thakur, K. Jha, K.C. and Sahi, B.P. (1986). Studies on the effects of different levels of nitrogen on the yield and quality of sugarcane ratoon. *Bharatiya Sug.* **11**(8): 39-42.

Thakur, K., Bose, P.K., Shahi, B.P. and Roysharma, R.P. (1979). Response of sugarcane to nitrogen and phosphate in calcareous soils. *Indian Sug.* **29**(5): 227-280.

Thanki, J.D., Patel, P.A. and Damame, H.S. (2000). Irrigation scheduling on the basis of climatological approach to sugarcane. *Indian Sug.* **49**(11): 905-907.

Thornthwaite, C.W. (1948). An approach towards a rational classification of climates. *Geog. Rev.* **38**:55-94.

Thyagarajan, C. (1982). The time and method of harvesting in ratoon management. *Seminar on the Ratoon management , Trichurapally, January, 1982.*

Toor, S.S., Singh,, Makhan, Garcha, A.I.S. and Singh, Sarjit (1999). Pre-*monsoon* irrigation requirement of spring planted sugarcane under subtropical conditions. *Indian Sug.* **49**(2) 1999): 129-133.

Vaidya, P.V. (1965). Further studies in the cultural and manural trials with sugarcane ratoon. *Proc. DSTA* **20**: 47-57.

Vaidya, P.V. and Kulkarni, N.G. (1962) Some preliminary trials on ratoon manuring. *Proc. DSTA* **18**: 142-148.

Vasudeva Rao, Naidu, R. and Lakshmikantham, M. (1946). Ratooning sugarcane in Madras. *Madras Agric. J.* **4**(1,2&3): 3-12.

Venkataraman, S., Gururaja Rao, P.N. and Naidu, K.M. (1986). The effects of water stress during the formative phase on stomatal resistance and leaf water potential and its relationship with yield in ten sugarcane varieties. *Field Crops Res.* **13**: 345-353.

Verma, H.P. (1962). Irrigation investigations conducted on sugarcane in India — A critical review. *Indian J. Sug. Cane Res. & Dev.* **6**: 76-79.

Verma, H.P. (1965). Studies on the inter-relation of levels of nitrogen and water duty in sugarcane varieties planted in spring. *The Indian Sugarcane J.* **9** (4): 219-222.

Verma, R.P. (1981). Extension strategy for increasing ratoon yield on farmers' fields. *Proc.National Seminar on Ratoon Management, Lucknow*. **II**: 213-217.

Verma, R.P. and Yadav, R.L. (1988). Transfer of ratoon management technology — a case study. *Indian J. Sugarcane Technol.* **5**(1): 50-53.

Verma, R.S. and Yadav, R.L. (1988). Intercropping in sugarcane for improving stubble sprouting under low temperature conditions in subtropical India. *Bharatiya Sug.* **13**(3): 45-48.

Verma, R.S., Panwar, B.S. and Srivastava, N.S.L. (1987). Effect of nitrogen doses and gap filling on ratoon of sugarcane varieties. *Coop. Sug.* **18**(9): 603-605.

Warner, H.L. and Leopold, A.C. (1969). Ethylene evolution from 2-chloro-ethyl phosphoric acid. Plant physiol. **44**: 56-58.

Winburne, J.N. (1962). "A Dictionary of Agricultural and Allied Terminology." Michigan State University Press, East Lansing, Michigan.

Wu, M.M., Kao, M.M., Chao, C. and Liu, C. (1975). The physiology and ecology of toxic codominant fungus in monoculture field in sugarcane. *Report of Taiwan Sugar Research Institute*, **68**: 31-42.

Yadav, D.V. and Yaduvanshi, N.P. S (1989). Micronutrients for increasing sugarcane production. *Indian Sug.***39**(4): 225-231.

Yadav, D.V. and Yaduvanshi, N.P.S. (1987). Micronutrients for increasing sugarcane production under limited water supply. *Proc. Ann. Conv. Pakistan Soc. Sugar Technol.* **23**: 325-332.

Yadav, D.V., Singh, Todi and Singh, Kishan (1987). Response of sugarcane to foliar application of micronutrients. *Indian J. Sugarcane Technol* **4**(1): 42-46.

Yadav, R.L. (1989). Production technologies for sugarcane under moisture stress and fertilizer constraints. *Bharatiya Sug.* 14 (10): 39-48

Yadav, R.L. (1992).Ratooning of sugarcane. Periodical Experts Book Agency, D-42, Vivek Vihar, Delhi, p.5.

Yadav, R.L. and Prasad, S.R., (1988). Moisture use

characteristics of sugarcane genotypes under different available soil moisture regimes in alluvial entisols. *J. Agric. Sci. Camb.* 110 (1): 5-11.

Yadav, R.L., Kumar, Ravi and Verma, R. S. (1990). Effect of nitrogen applied through new carriers on yield and quality of sugarcane. *J. Agric. Sci. Camb.* **114**(2): 225-300.

Zende, G.K. (1973). Sugarcane News 6:21.

Additional References

Agrawal, D.C. and Gupta, R.A. (1960). Improvement in the practices for ratooning of sugarcane — A review. *Indian Sug.* **10**(12): 847.

Agrawal, M.L, Nath, A., Singh,R. and Ali, I. (1976). Research work on sugarcane ratoon in Uttar. Pradesh. *Proc. Seminar on Ratoon Management, STAI, April 26, 1976, Poona.*

Agrawal, M.L. (1981). Sugarcane ratooning studies in Uttar Pradesh — A review. *Proc. National Seminar on Ratoon Management, Lucknow.* **III**: 242-245.

Agrawal, M.L., Bharadwaj, K.M., Singh, R. and Pathak, V.K. (1976). Effect of nitrogen and its time of application on yield and quality of ratoon. *Proc. Seminar on Ratoon Management, STAI, April* **26**, 1976, Poona. pp. 41-48.

Agrawal, M.P. and Singh, M. (1986). Effect of mulches on soil temperature and sprouting of sugarcane ratoons. *International J. Trop. Agric.* **4**(1): 23-29.

Agrawal, M.P. and Srivastava, A.K. (1983). Sprouting in sugarcane ratoons under variable temperature conditions. *Proc. Ann. Conv. STAI* **47**: Ag 73-77.

Ahmad S., Nasir and Giridharan, S. (2000). Study on the influence of management practices on sugarcane ratoon crop. *Indian Sug.* **49**(11): 835-838.

Anonymous (1940-41). Comparison of organic and inorganic manures and their mixtures. *Ann. Rep. SRS. Muzaffarnagar*, p. 87.

Anonymous (1941). Best time of harvesting plant cane for keeping ratoons. *Ann. Rep. Sug. Res. & Dev. Workers in U.P. (Muzaffarnagar, 1940-41)*, p.103.

Anonymous (1941). Residual effects of ratoons from the experiment with *sanai* and cotton rotation. *Ann. Rep. Sug. Res. Workers in U.P. (Muzaffarnagar, 1940-41). p.96.*

Anonymous (1941). To determine the period of profitable ratooning of cane. *Ann Rep. Sug. Res. Workers in U.P. (Shahjahanpur, 1940-41), p. 3*

Anonymous (1992). Multiple ratooning: Farmer Profulla Chandra excels. *Indian Sug.* **42**(5): 285.

Arnata, R. (1968). What makes a good ratoon crop? *Sugarland (Philippines)* **5**(7): 19, 32.

Balbad, S.G. (1951). Fap filling in ratooning. *Proc. DSTA.* **8**: 117-120.

Baliyan, S.P. and Bhogal, T.S. (1996). Resource use efficiency in ratoon crop of sugarcane in western Uttar Pradesh. *Indian Sug.* **45**(12): 943-950.

Behal, K.L. (1964). Juice quality of sugarcane ratoon as affected by different dates of harvesting and application of irrigation to the preceding plant crop during hot months. *Proc. Bien. Conf. Sug. Res. & Dev. Workers* **5**: 62-70.

Bhadauria, V.S. (1968). Study on the nitrogen requirement of sugarcane ratoon crop in the eastern tract of U.P. *Indian Sug.* **17**(11): 809-814.

Bhale, V.M. (1996). Effect of growth regulators on ratoon productivity of cane cultivars. *Proc. Ann. Conv. STAI* **58**: 67-74.

Bhale, V.M. and Hunsigi, G. (1994). Effect of growth regulators and cultural treatment on productivity of ratoon cane. *Indian Sug.* **44**(8): 645-652.

Bhardwaj, K.M., Singh, M., Rastogi, R.P. and Sharma, R.K. (1977) Effect of nitrogen doses on sugarcane plant and ratoon crops and *gur* production. *Sugar News* **8**(12): 7-10.

Bhatt, S. and Shukla, S.P. (1989). Comparative juice qualities of fall planted, spring planted and ratoon sugarcane crops at different harvesting periods. *Proc. Ann. Conv. STAI* **52**: Ag 137-150.

Bhoj, R.L. (1971). Improvement in the practices of ratooning in sugarcane. *Indian Sug.* **20**(12): 847-853.

Bhuta, J.C. and Parasar, K.S. (1981). Studies on the effect of intercropping and N fertilizer on the yield and quality of sugarcane first ratoon. *Indian Sug.* **31**(4): 272-279.

Bindhumadhava Rao, R.S. and Venu Reddy, R. (1982). Ratoon in sugarcane: Some aspects. *Proc. Seminar on Ratoon Management, Trichanapally, January, 1982.*

Boramanikar, P.K., Patil, S.A. and Medhane, N.S. (1993). Effect of direct incorporation of sugarcane trash in preseasonal and ratoon crop and its effect on yield and quality of sugarcane. *Indian Sug.* **42**(12): 913-916.

Brave, D.B. (1950). Ratoon crop and its quality. *Proc. DSTA.* **17**: 1-4.

Bugragohain, S.K. and Singh, S.N. (1993). Gap filling-cum-fertilizer management in ratoon cane. *Indian J. Sugarcane Technol.* **8**(1): 79-86.

Chaudhary, D.C., Singh, S.N. and Singh, S.B. (1992). Studies on the effect of time and fractional doses of nitrogen application on yield, yield attributes, juice quality and economic return of sugarcane ratoon. *Coop. Sug.*

24(2&3): 99-102.

Chaudhary, S.K. (1998). Ratooning ability of early maturing sugarcane genotypes under low management conditions. *Indian J. Sugarcane Technol.* **13**(2): 57-61.

Chauhan, R.S. (1993). Effect of irrigation schedule and nitrogen levels on late started ratoon crop of sugarcane *Bharatiya Sug.* **18**(5): 57-58.

Chauhan, R.S. and Bajpai, P.K. (1989). Effect of cane lodging on growth yield and quality of sugarcane ratoon. *Bharatiya Sug.* **15**(1): 155-157.

Chauhan, R.S. and Chandra, J. (1991). Gaps in sugarcane ratoons — A review. *Coop. Sug.* **22**(10): 655-658.

Chauhan, R.S. and Yadav, R.L. (1988). Intercropping of green gram in plant crop of sugarcane with residual effect on subsequent ratoon crop. *Coop. Sug.* **20**(5): 315-317.

Chauhan, R.S., Yadav, R.L. and Bajpai, K.K. (1991). Effect of N application with and without manganese on yield of ratoon cane. *Bharatiya Sug.* **16**(4): 59.

Cheema, S.S. and Singh, Karnail (1970). Better ratooning– A way to improving cane yield. *Coop. Sug.* **1**(10): 499-500.

Chitkala Devi, T., Rao, K.R., Swamy, K.R.L. and Raju, D.V.N. (1997). Influence of irrigation and fertilizer levels on ratoon yield and quality of Co 7219 sugarcane. *Proc. Ann. Conv. DSTA.* **46**(1): 131-140.

Chougle, J.D. (1979). Aerial spray of nitrogen (urea) to increase the yield of sugarcane ratoon. *Maharashtra Sug.* **4**(5): 41-42.

Chougule, J.D. (1977). Optimum dose of NPK fertilizer for ratooning of sugarcane under deep black soils of Maharashtra. *Indian Sug. Crops J.* **4**(2): 49.

Chougule, J.D. (1979). Cultivation of sugarcane ratoon is economic than plant cane. *SISTA Sugar J.* **5**(2): 25-27.

Chougule, J.D. and Patil, B.R. (1982). Cultivation of sugarcane ratoon is economic than plant cane in medium black soil of Maharashtra. *Maharashtra Sug.* **7**(6): 43-45.

Chougule, J.D. and Patil, K.D. (1977). Effect of intercultivation (by bullock pairs) on the growth and yield of sugarcane ratoon in black soils. *Indian Sug. Crops J.* **4**(2): 45-46.

Chougule, J.D. and Patil, K.D. (1979). Intercultivation increases the yield of sugarcane ratoon in deep black soils. *Maharashtra Sug.* **4**(7): 21-22.

Chougule, J.D. and Wandre, S.S. (1981). Proper maintenance of good sugarcane ratoon. *Maharashtra Sug.* **6**(6): 39-40.

Chugare, R.V., Bankar, M.C., Mane, T.A. and Kenjale, S.T. (1994). Effect of drip and surface methods of irrigation on yield and quality of ratoon sugarcane Cv. Co 7219. *Indian Sug.* **44**(7): 503-506.

Chugare, R.V., More, V.D. and Kenjale, S.T. (1995). Effect of levels of irrigation and nitrogen on yield and quality of ratoon sugarcane (Co 7219). *Bharatiya Sug.* **21**(8): 7-9.

Chui, P. and Samuale, G. (1977). Evaluation of nitrogen fertilizer efficiency for plant cane and ratoon crops in irrigated sugarcane. *Proc. ISSCT.* **16**: 1259-1264.

Dakshindas, D.G. (1976). Sugarcane ratoon — A lead paper. *Proc. Seminar on Ratoon Management, ASTI, April 26,1976, Poona, pp.* 1-13.

Daniel, K.V., Sugumaran, T. and Kumar, Suresh (1976). Effect of nitrogen and potassium on the yield and quality of sugarcane ratoon. *Coop. Sug.* **7**(7): 341-346.

Dey, P. and Yadav, D.V. (1996). Chlorophyll scan, yield and juice quality of chlorotic sugarcane ratoon as influenced by foliar application of ferrous sulphate. *Indian J. Sugarcane Technol.* **11**(2): 131-135.

Dhilon, P.S. and Singh. Nazar (1981). Ratoon management of sugarcane in Punjab. *Indian Sug. Crops. J.* **8**(2): 23-25.

Divakar, M.V (1987). Management of ratoons — A review. *Proc. DSTA* **37**: A139-143.

Durai, R. (1998). Different methods of gap filling in sugarcane ratoon. *Indian J. Agron.* **43**(2): 346-347.

Durai, R. and Devaraj (1994). Study of the response of different sugarcane varieties under ratoon condition to graded doses of nitrogen. *Bharatiya Sug.* **20**(9): 29-31.

Durai, R. and Devaraj, G. (1994). Study of the response of different sugarcane varieties under ratoon condition to graded doses of nitrogen. *Bharatiya Sug.* **20**(9): 29-31.

Durai, R., Ahmad S., Nasir., Pandian, B.J. and Rajasekaran, S. (1989). Studies on the effect of spatial variations on yield and quality of sugarcane in plant and ratoon crop. *Bharatiya Sug.* **14**(12): 37-39.

Durai, R., Ahmed, M. and Rangiah, P.K. (1988). Studies on ratoon management in sugarcane. *Bharatiya Sug.* **13**(11): 81-82.

Ekambaram, C. (1949). Some aspects of ratooning in sugarcane. *Madras Agric. J.* **6**(1): 28-29.

Ethirajan, A.S. (1982). Varieties in relation to ratoon

management. *Paper presented at the SBI R&D Workers meeting, Pettavathalai, Trichy.*

Ethirajan, A.S., Shahi, B.K. and Misra, A. (1976). Effect of certain management practices on the performance of the ratoon of Co 419. *Sug. Cane Semin. Mandya*, 11-13.

Gholve, S.G., Bhoite, D.S. and Kumbhar, S.G. (1997). Trash management in ratoon sugarcane and its effect on cane yield and soil fertility. *Bharatiya Sug.* **23**(1): 107-112.

Ghosh, A.K., Shrivastava, A.K. and Saxena, Y.R. (1987). Rooting pattern and nutrients depletion studies in plant and ratoon crops of sugarcane. *Indian J. Sugarcane Technol* **4**(1): 47-52.

Gulati, J.M.L., Misra, M.M., Bahera, A.K. and Hati, N. (1993). Performance of sugarcane (ratoon) under constraints of water and nitrogen. *J. Water Management* **1**(2): 102.

Gupta, A.P. and Prasad, B. (1971). Studies on the composition of cane juice II :Effect of nitrogen manuring on various inorganic non-sugars in cane juice of plant and ratoon crops. *Proc. Jt. Conv. STAI, DSTA and SISTA, Poona,* **4**: Ag 91-98.

Gupta, A.P. and Prasad, B. (1971). Studies on the composition of cane juice I: Effect of nitrogen manuring on various inorganic non-sugars in cane juice of plant and ratoon crops. *Proc. Jt. Conv. STAI, DSTA and SISTA, Poona,* **4**: Ag 81-90.

Gupta, R.A. (1963). On efficiency of seed cane obtained from spring, autumn and ratoon crops for planting in autumn. *Indian J. Sug. Res. & Dev.* **8**(1): 81-83.

Hora, B.S. and Motiwale, M.P. (1981). Raising multiple ratoons in north India. *Proc. National Seminar on*

Ratoon Management, Lucknow. **II**: 154-157.

Hsia, Y.J. and Yang, C. U. (1968). Studies on poor ratooning of cane in Taiwan. *Proc. ISSCT* **13**: 623-634.

Jafri, S.M.H. (1984). Effect of residual phosphate and potash on multiple ratoons of sugarcane. *Proc. Ann. Conv. STAI* **48**: Ag. 183-191.

Jafri, S.M.H. (1987). Effect of potassium with nitrogen and phosphorus on sugarcane in plant ratoon cropping sequence in an alluvial soil. *J. Indian Soc. Soil Sci.* **35**(4): 667-671.

Jafri, S.M.H. and Alam, M. (1981). Nutritional requirements of ratoon crops. *Proc. National seminar on Ratoon Management, Lucknow.* **I**: 60-71.

Jafri, S.M.H. and Pandit, S.N. (1987). Nitrogen use efficiency and release through legume in plant ratoon-ratoon cropping sequence. *Indian J. Sugarcane Technol* **4**(1): 86-89.

Jafri, S.M.H., Singh, T. and Ghosh, A.K. (1991). Effect of N and K application in sugarcane grown in plant-ratoon cropping sugarcane on an alluvial soil. *J. Potassium Research* **7**: 198-206.

Jayabal, V. and Chockalingam, S. (1991). Studies on the improvement of ratoon yield of sugarcane variety CoC 671. *Coop. Sug.* **23**(4): 251-254.

Kamlesh, K. and Kanwar, R.S. (1986). Ratooning behaviour of new promising varieties of sugarcane under low temperature conditions. *Bharatiya Sug.* **11**(5): 47-48.

Kanwar, R.S. (1977). Improved cane cultivation techniques and management of ratoon crop. *Sugar News* **8**(12): 15-18.

Kanwar, R.S., Garcha, A.I.S., Singh, Sarjit and Bains, B.S.

(1989). Effect of hot air treatment, interrow spacing and sett size on yield and quality of sugarcane ratoon crop. *Indian Sug.* **39**(3): 175-178.

Kapoor, M.L. (1993). Effect of source and level of nitrogen application on the yield and quality of plant and ratoon crops of sugarcane. *Ferti. News* **38**(11): 27-31.

Kapur, R., Bhat, S.R. and Tripathi, B.K. (1985). Ratooning ability of sugarcane genotypes under late harvested conditions. *Indian Sug.* **35**: 445-448.

Khudapur, G.J. (1983). Ratooning of sugarcane in India. *Maharashtra Sug.* **8**(4): 55-58.

Kumar, A. and Pandey, H.P. (1991). Effect of late application of growth regulators on sugars and hydrolytic enzymes of sugarcane stubble bonds and subsequent effect on ratoon regrowth. *Coop. Sug.* **22**(7): 449-453.

Kumar, M. Dinesh, Hunsigi, G. and Shankaraiah, C. (1996). Effect of time and source of nitrogen application on yield and quality of ratoon cane. *Madras Agric. J.* **83**(7): 433-435.

Kumar, Ravi (1985). Exploring the possibility of harvest of plant cane above the ground level to improve the yield and quality of sugarcane ratoon crops—A new approach in ratoon management. *Indian Sug.* **34**(11): 75-76.

Kumar, Ravi, Verma, R.S. and Yadav, R.L. (1986). Nutrient requirements of sugarcane ratoon crop in north India. *Coop. Sug.* **17**(10): 779-983.

Lal, J., Singh, K. and Gupta, R.K. (1981). Trends and variability in area, production and productivity of plant and ratoon crops of sugarcane in Uttar Pradesh. *Proc. National Seminar on Ratoon Management, Lucknow.*

II: 162-169.

Lal, Menhi, Shukla, S.K. and Jaiswal, V.P. (1997). Management of sugarcane ratoon. *Sugar Crops Newsletter* **7**(1&2): 2-3.

Mahendran, S., Karamathullah, J., Porpavai, S. and Ayyam, A. Perumal (1995). Effect of planting systems and ratoon management on the yield and quality of ratoon crop. *Bharatiya Sug.* **22**(1): 123-127.

Mahendran,S., Duraismy K.S. and A., Ayyamperumal (1993). Effect of nitrogen on flowering behaviour, yield and quality of ratoon sugarcane cultivation under Cauvery delta zone. *Bharatiya Sug.* **18**(11): 39-40.

Manisegaran, S., Rechiappan, S., Gopalan, M. Purushothaman, S. and Ramiah, S. (1988). Effect of irrigation regimes, method of irrigation and fertilizer application on the incidence of early shoot borer in ratoon sugarcane. *Indian Sug.* **38**(6): 471-474.

Mathan, K.K. (1998). Effect of cytozyme on ratoon cane yield (Var. Co 86010). *Coop. Sug.* **29**(6): 405-408.

Mathur, B.S. (1977). Ratooning in sugarcane. *Coop. Sug.* **8**(7): 351-354.

Mathur, P.S. and Misra, A. (1984). On improving the productivity of sugarcane ratoon taken from winter and late harvested plant crops. *Indian Sugar Crops. J.* **10**(3): 1-4.

Mathur, P.S. and Seksena, M.M. (1965). On the utility of trash mulch in sugarcane ratoon. *Indian J. Sug. Res. & Dev.* **10**(1): 24-27.

Mehrotra, O.N. (1981). Some suggestions about physiological research on ratoon cropping of sugarcane. *Proc. National Seminar on Ratoon Management, Lucknow.*

III: 240-241.

Misra, A. and Mathur, P.S. (1986). Effect of clump density and fertility levels on yield and quality of ratoon raised from winter harvested sugarcane. *Indian J. Agron.* **31**(2): 172-176.

Misra, A. and Mathur, P.S. (1991). Ratooning of sugarcane varieties as influenced by time of harvest and facility levels. *Indian J. Sugarcane Technol.* **6**(2): 1-7.

Motiwale, M.P. and Alam, M. (1981). Certain agronomic aspects of ratoon management. *Proc. National Seminar on Ratoon Management. Lucknow.* **I**: 96-99.

Nagar, D.S. (1948). Effect of nitrogen application on ratoon crops of sugarcane. Proc. *Ann. Conv. STAI* **16**(II): 18-24.

Naidu, N.V., Raja Bapa Rao, V. And Dora, K.B. (1997). Performance of early maturing varieties in ratoon crop. *Indian Sug.* **46**(11): 871-875.

Nath, A. and Mathur, B.K. (1965). Time of manuring ratoon under deficient irrigation conditions. *Indian Sug.* **15**(12): 803-806.

Nayak, N., Jena, B.C. and Mahapatra, S.S. (1998). Ratoon management in sugarcane. *Indian Sug.* **48**(7): 501-504.

Neelakantham, P., Kurian, S.and Chandy, K.C. (1991). Investigation on ratoon decline in sugarcane. IV Path coefficient analysis in plant and ratoon sugarcane. *Indian J. Sugarcane Technol.* **6**(2): 103-108.

Nevase, V.B., Talathi, P.G., Wagh, R.G. and Khan Vilkar, S.A. (1994). Ratoon performance of three sugarcane genotypes under the nitrogen levels in Konkani region. *Bharatiya Sug.* **20**(8): 47-49.

Nigade, R.D., Ankalkoppe, M.N. and Jadhav, B.S. (1999).

Management of sugarcane ratoon. *Bharatiya Sug.* **24**(3): 9-14

Pandey, M.B., Chauhan, R.S., Srivastava, T.K., Yadav, R.L. and Singh, G.B. (1997). Effect of time of ratoon initiation and earthing up on growth and yield of sugarcane ratoon. *Indian J. Sugarcane Technol.* **12**(1): 81-83.

Pandit, S.N. and Sinha, A.K. (1981). Increasing fertilizer efficiency in sugarcane ratoon. *Proc. DSTA* **31**: 97-102.

Paneerselvan, R. (1991). Effect of management practices and N levels on sheath moisture per cent at different stages of sugarcane ratoon. *Bharatiya Sug.* **17**(4): 47-49.

Paneerselvan, R., Jayabal, V., Tamilselvan, N. and Sankaran, N. (1991). Effect of growth regulators, nutrient solution and N levels on yield and yield attributes of sugarcane ratoon. *Indian Sug.* **41**(7): 543-546.

Pannerselvan, R., Sankaran, N. and Jayabal, V. (1991).Effect of integrated management practices on sprouting and growth characters of sugarcane ratoon. *Bharatiya Sug.* **16**(9): 23-27.

Panwar, B.S. (1996). Enhancing ratoon yields by proper gap filling and trash mulching in sugarcane. *Bharatiya Sug.* **21**(10): 49-52.

Panwar, B.S. and Singh, B.V. (1997). Effect of gap filling-cum-fertilizer management in ratoon crop of sugarcane. *Bharatiya Sug.* **23(**1): 77-78.

Panwar, B.S., Verma, R.S. and Srivastava, S.N.L., (1989). Effect of cultural practices and irrigation on winter started ratoon crop of sugarcane. *Coop. Sug.* **20**(6): 409-410.

Parashar, K.S., Prasad, A. (1985). Studies on various

intercropping and management practices on yield attributes, yield and juice quality of first sugarcane ratoon. *Indian Sug.* **35**(2): 121-125.

Parekh, S.K. (1966). Ratooning of sugarcane in Bihar. *Indian Sug.* **15**(10): 671-674.

Patel, H.S., Shah, T.M., Arvadia, M.K. and Patel, M.P. (1988). Effect of gap filling and nitrogen levels in yield and quality of ratoon raised sugarcane varieties. *Proc. DSTA* **38**(I): A165-170.

Patil, A.S. and Zende, G.K. (1983). Effect of soil salinity on tissue indices and yield of ratoon sugarcane. *Maharashtra Sug.* **8**(4): 33-36.

Patil, V.B., Jade, B.D. and Karwa, A.N. (1996). Improved cultural practices for sugarcane under micro (Drip) irrigation systems in Maharashtra. *Proc. Ann. Conv. STAI* **58**: 123-128.

Pawar, B.S. and Singh, B.V. (1994). Enhancing ratoon yields by proper crop management. *Coop. Sug.* **25**(9-10): 373-374.

Pawar, H.K. and Patil, B.R. (1989). Studies on the effect of layouts, earthing up and irrigation schedules to *adsali* sugarcane on growth and yield of its ratoon with different schedules of irrigation. *Bharatiya Sug.* **14**(12): 27-30.

Peng, S.Y. and Twu, L.T. (1978). Application of plant growth substances to improve germination and yield of ratoon cane. *Taiwan Sugar* **25**(1): 8-17.

Pol, K.M. and Deshmukh, S.B. (1999). Comparative physiological studies of seasonal plant and ratoon crops of six sugarcane varieties II. Co 7527 and CoM 7125. *Bharatiya Sug.* **25**(2): 7-10.

Potty, N.N., Kurien, S. and Chandy, K.C. (1987).

Investigations on the phenomenon of ratoon decline in sugarcane. *Proc. Ann. Conv. STAI* **50**: Ag 33-40.

Prasad, B., Singh, K.D.N., Prasad, C.R. and Prasad, J. (1985). Fertilizer recommendation for economic yield of ratoon sugarcane based on soil test values in calcareous soil of Bihar. *Indian Sug. Crops J.* **11**(1-14): 39-45.

Prasad, S.R., Sharma, R.K. and Alam, M. (1981). Yield of ratoon in relation to intensity of gaps. *Proc. National Seminar on Ratoon Management, Lucknow*. **I**: 72-74.

Rai, R.K., Sharma, C.P., Dube, B.K., Motiwale, M.P. and Singh, A.B. (1999). Improvement of yield and quality of a ratoon crop of sugarcane variety CoLk 8102 through organic amendments in non-saline sodic soil. *Coop. Sug.* **30**(10&11): 1047-1053.

Raju, J.S.N., Jaganamohan Rao, R., Subramanyeswara Rao, C.V,. Paparao, J., Gopala, B and Prasada Rao, K.K. (1981). Studies on the effects of certain cultural practices on the performance of sugarcane ratoons. *Maharashtra Sug.* **6**(8): 9-22.

Ramakrishna Rao, S. and Veerabhadra Rao, K. (1988). Studies on micro-nutrients composition of certain soils and plant and ratoon cane crops raised on the same field in Pithapuram factory area of A.P. *Bharatiya Sug.* **13**(5): 45-50.

Ramesh, P. and Mahadevaswamy, M. (1996). Effect of planting season and harvesting age of plant and ratoon crops on yield and quality of sugarcane varieties. *India J. Agric. Sci.* **66**(11): 641-644.

Rao, K.L., Chitkala Devi, T. and Raju, D.V.N. (1997). Effect of time of ratooning and age of harvest on yield and quality of early, mid and late sugarcane varieties. *Proc. Ann. Conv. DSTA* **46**(I): 141-150.

Rao, K.L., Raju, D.V.N. and Chitkala Dev, T. (1994). Effect of N and K nutrition on ratoon cane crop raised under rainfed condition. *Proc. Ann. Conv. DSTA* **44**: 41-54.

Rao, K.L., Raju, D.V.N. and Chitkala Devi, T. (1995). Effect of N and K nutrition on ratoon cane crop raised under rainfed condition. *Bharatiya Sug.* **22**(1): 55-64.

Rathi, K.S. and Singh, R.A. (1979). Effect of companion cropping sequences on the yield and quality of sugarcane ratoon. *Indian Sug.* **29**(1): 15-18.

Sachan, R.S., Ram, Nand and Gupta, R.A. (1993). Effect of soil and applied nitrogen, phosphorus and potassium on the yield of planted and ratoon crop of sugarcane in a mollisol of U.P. *Indian Sug.* **42**(10): 774-769.

Selvam, R.P., Jayabal, V., Tamilselvan, N. and Sankaran, N. (1991). Studies on ratoon management in sugarcane. *Bharatiya Sug.* **16**(11): 27-30.

Sen, A. Singh, A.P. and Prasad, C.R. (1986). Effect of soil application of zinc on plant ratoon system of sugarcane grown in calcareous soil. *Proc. Ann. Conv. STAI* **49**: Ag 146-152.

Sharma, S., Uppal, S.K. and Kalsi, P.S. (1999). Response of terpenoids on ratoonability, growth and quality of sugarcane. *Indian Sug.* **49**(4): 283-286.

Sharma, S.C. and Gupta, M.L. (1991). Juice quality of sugarcane ratoon as infuenced by nitrogen sources, levels and time of application. *Coop. Sug.* **22**(9): 592.

Sharma, S.N., Singh, S., Sharma, R.P., Parashar, K.S. and Prasad, R. (1981). Studies on nitrogen carry over effects of urea, sulphur coated urea, *neem* cake coated urea and nitropyrine treated urea applied to sugarcane plant crop on succeeding ratoon. *Z. Acker Und plan-zenban.*

150: 207-214.

Shinde, D.B., Navale, A.M. and Jadhav, S.B. (1993). A study on incorporation of sugarcane trash in sugarcane ratoon. *Journal of Maharashta Agricultural Univeristy* **18**(2): 264-265.

Shrivastava, A.K., Ghosh, A.K. and Agnihotri, V.P. (1992). **Sugarcane Ratoons**, Oxford and IBH Publishing Co. Pvt. Ltd., 66 Janpath, New Delhi.

Shrivastava, A.K., Singh, K. and Kacker, N.K. (1981). Partitioning of dry matter, energy conservation efficiency and energy productivity in plant canes and ratoons. *Proc. National Seminar on Ratoon Management, Lucknow.* **I**: 82-91.

Shrivastava, A.K., Yadav, R.L., Singh. K., Kacker, N.K. and Singh. M. (1981).Stubble sprouting in sugarcane under subtropical conditons: Influence of growth regulating substances. *Proc. National Seminar on Ratoon Management, Lucknow.* **I**: 100-108.

Shrivatava, A.K., Singh, M. and Bajpai, P.K. (1982). Influence of soil nitrogen and organic matter on dynamics of tiller production in sugarcane ratoon. *Indian J. Pl. nutrition.* **1**: 107-113.

Shukla, U.S., Singh, V. and Verrna, R.P. (1989). Recent advances in management of sugarcane ratoons. *Coop. Sug.* **21**(3): 183-186.

Siag, R.K., Verma, B.L. and Deo, R. (1989). Effect of different levels of irrigation and nitrogen fertilizer on the yield and quality of plant and ratoon cane crops in north west Rajasthan. *Proc. Ann. Conn. STAI* **52**: Ag. 39-46.

Siag, R.K., Verma, B.L. and Deo, R. (1991). Effect of irrigation scheduling and nitrogen levels on water use, yield quality of sugarcane ratoon. *Madras Agric. J.* **78**(9-12): 444-449.

Singh, A.K., Lal, Menhi and Shukla, S.K. (1998). Ratooning in sugarcane : Present and future. *Bharatiya Sug.* **23**(6): 29-32.

Singh, G.B. and Sharma, R.K. (1975). Efficiency of levels and time of nitrogen application in sugarcane ratoon. *Ann. Rep. IISR, Lucknow*, pp. 15-16.

Singh, G.B. and Sharma, R.K. (1975). Levels and time of nitrogen application for sugarcane ratoon. *Ann. Rep.IISR, Lucknow*, pp. 25-26.

Singh, H.N., Singh, G.B. and Verma, P.S. (1985). Selection parameter for sugarcane ratoon. *Coop Sug.* **16**(6): 339-341.

Singh, Kishan (1983). An approach to sugarcane ratoon development in Captainganj (Deoria) U.P. *Indian Sug.* **32**(12): 929-934.

Singh, Kishan (1996). Multiple ratoon management in sub-tropical India. *Indian Sug.* **46**(5): 325-328.

Singh, M. and Agrawal, M.P. (1985). Nitrogen transformation during sugarcane root decomposition. *J. Indian Sec. Soil Sci.* **33**: 676-678.

Singh, N.L., Pandey, B.N., Gangasaran and Dayal R. (1973). Effect of trash burning on plant growth and pest incidence in sugarcane ratoon during hot weather. *Sugar News* 5: 18-20.

Singh, N.P., Rai, S.P. and Singh, R.B. (1999). Effect of chemical and cultural treatments on sprouting of stubble buds in winter harvested plant crop of sugarcane. *Coop. Sug.* **31**(3): 203-206.

Singh, O. and Kanwar, R.S. (1988). Combined effect of ethrel and lindane on low temperature ratooning of sugarcane. *Coop. Sug.* **19**(6): 381-385.

Singh, O. and Kanwar, R.S. (1988). Ratoon of cold harvested sugarcane as influenced by exogenous application of ethephon. *Indian J. Sugarcane Technol.* **5**(1): 69-71.

Singh, Omkar (1973). Studies on the response of ratoon of sugarcane to calcium ammonium nitrate. *Indian Sug.* **22**(10): 781-784.

Singh, Onkar (1995). Effect of terpenoids on the sugarcane ratoon of cold harvested crop of early maturing variety. *Coop. Sug.* **26**(9): 701-704.

Singh, P.P. (1974). Ratooning potential of sugarcane varieties under different levels of nitrogen. *Proc. Ann. Conv. STAI*: **40**(I): 1-5.

Singh, P.P. (1975). Effect of harvesting dates on the yield and quantity of plant and ratoon crops of sugarcane. *Proc. Joint Conv. STAI, DSTA & SISSTA* **5**: Ag 11-16.

Singh, P.P. (1975). Effect of harvesting dates on the yield and quality of plant and ratoon crops of sugarcane. *Proc. Ann. Conv. STAI*: 11-16.

Singh, S. (1981). The optimum and suitable methods of harvesting plant crop for higher yield in ratoon. *Paper Presented in 3rd SBI Science Club in ratoon management at Cauvery Sugar Factory, 1981.*

Singh, S.N., Singh, S.B., Singh, N.P. and Chaudhary, D.C. (1995). Studies on the effect of plant cane harvesting and time of nitrogen application on yield and quality of sugarcane ratoon. *Coop. Sug.* **26**(5): 359-364.

Singh, T., Yadav, D.V. and Srivastava, A.K. (1988). Comparative evaluation of soil and foliar applied nitrogen with insecticide in sugarcane ratoon. *Indian J.*

Sugarcane Tehnol **5**(1) 66-68.

Sinha, U.P., Jha, K.C., Dubey, B. and Kumar, S. (1995). Effect of agronomic management practices on the yield and quality of ratoon sugarcane. *Bharatiya Sug.* **21**(6): 33-35.

Soopramanian, G.C. and Hunsigi, G. (1996). Ratooning, nutrition and environment : An overview. *Coop. Sug.* **27**(11): 831-850.

Sree Kumar, S.G., Gopinath Nair, V., Saraswathey, P. and Sukumaran Nair, S. (1991). Association among cane yield, sugar yield and other components in the first ratoon crop of sugarcane. *Bharatiya Sug.* **16**(11): 53-56.

Sundara, B. (1992). Varietal characteristics affecting ratooning potential of sugarcane. *Sugar Cane* **6**: 1-4.

Sundara, B. (1996). Studies on multiratooning. *Proc. Ann. Conv. STAI* **58**: 3-12.

Sundara, B. (1997). Multiratooning potential of sugarcane varieties and management of multiratoons. *Indian J. Sugarcane Technol.* **12**(1): 84-89.

Sundara, B. and Tripathi, B.K. (1999). Available N changes and N balance under multi-ratooning of sugarcane varieties in a tropical vertisol. *Proc. ISSCT* **XXIII** (II): 80-88.

Tiwari, R. and Dheer, D.P. (1994). Resource use efficiency of ratoon. *Indian Sug.* **44**(2): 109-112.

Tomar, P.S. , Mathur. O.P. and Oberai, D.S. (1965). Iron deficiency on ratoon crop of sugarcane in canal irrigated soil of Rajasthan. *Indian J. Sug. Res. & Dev.* **9**(2): 123.

Tripathi, B.K. and Gill, S.S. (1987). Genetic evaluation of varietal crosses in sugarcane (*Saccharum* spp. complex) for economic trials in ratoon crop. *Indian J. Sugarcane Technol.* **4**(10): 61-65.

Tripathi, B.K., Gill, S.S., Misra, G.P. and Lal, S. (1981). Screening of sugarcane (*Saccharum* species complex) genotypes for ratooning ability. *Proc. National Seminar on Ratoon Management, Lucknow.* **II**: 206-212.

Tripathi, B.K., Gill, S.S., Misra, G.P. and Lal, S. (1982). Screening of sugarcane (*Saccharum* hybrids) genotypes for ratooning ability. *Indian Sug.* **32**(9): 577-581.

Vaidya, B.R., Kadam, B.S., Akolkar, R.D. and Jadhav, S.B. (1992). Effect of foliar application of micro-nutrient formulations on the yield and quality of ratoon (Co 740). *Proc. DSTA* **42**: A 227-231.

Vaidya, P.V. and Patel, J.M. (1963). Studies on cultural and manural trials with sugarcane ratoons. *Proc. Ann. Conv. STAI* **31** (I): 15-32.

Veerabhadra Rao, K., Ramakrishna Rao, S., Ramalingaswamy, K. and Bapu, N. (1987). Studies on micro-nutrient composition of soils and ratoon cane crops raised on the same fields in Anakapalle, Vuyyuru, Govada, Payakaraopeta and Bhimasinghi factory area of A.P. *Bharatiya Sug.* **12**(5): 57-64.

Venugopal, M.R. (1994). Sugarcane ratoon. First culture is agriculture, sugarcane culture is sweet culture. *Indian Sug.* **44**(2): 119-121.

Verma, K.S. and Dang, Y.P. (1995). Phosphorus requirement of sugarcane ratoon crop. *Coop. Sug.* **26**(8): 623-628.

Verma, R.S. (1995). Ratoon decline in sugarcane. *Coop. Sug.* **26**(5): 349-351.

Verma, R.S. and Yadav, R.L. (1988). Intercropping in sugarcane for improving stubble sprouting under low temperature conditions in subtropical India. *Bharatiya Sug.* **13**(3): 45-48.

Verma, R.S., Kumar, A. and Prakash, O. (1993). Proper cane ratoon management for high returns. *Coop Sug.* **24**(6): 243-244.

Yadav, R.L. (1979). Ratoon management in sugarcane — An agronomic prospective. *Indian Sug. Crops J.* **6**(4): 83-86.

Yadav, R.L. (1983). Dry matter production, yields and nutrient uptake by sugarcane ratoon as affected by methods of planting and earthing up. *Indian J. Agron.* **28**(3): 214-217.

Yadav, R.L. (1983). Yield and quality of sugarcane ratoon as influenced residual and direct N application. *Indian Sugar Crops. J.* **9**(3): 15-17.

Yadav, R.L. (1986). Bumper ratoons from winter harvested sugarcane. *Indian Fmg.* **36**(6): 21-22.

Yadav, R.L. (1986). Sustaining ratoon cane yields with efficient N management practices. *Ferti. News.* **31**(6): 17-22.

Yadav, R.L. (1992). **Ratooning of sugarcane**. Periodical Experts Book Agency, D-42, Vivek Vihar, Delhi.

Yadav, R.L. and Prasad, S.R. (1991). Residual effect of soil and foliar applied potassium on rainfed plant crop of sugarcane to its ratoon. *Bharatiya Sug.* **16**(9): 59-61.

Yadav, R.L. and Sharma, R.K. (1983). Effect of nitrogen application to the plant crop of sugarcane on the yield and quality of the subsequent ratoon crop. *Indian J. Agric. Sci.* **53**(1): 38-43.

Yadav, R.L., Prasad, S.R., Singh, R. and Srivastava, V.K. (1994). Recycling sugarcane trash to conserve soil organic carbon for sustaining yields of successive ratoon crops in sugarcane. *Bioresource Technol.* **49**: 231-235.

Yaduvanshi, N.P.S. and Yadav, D.V. (1993). Effects of combined application of sulphitation pressmud and fertilizer nitrogen on the yield, quality and nitrogen economy in sugarcane ratoon. *Natl. Seminar on soil Resource vis-à-vis Sustainable sand Use at Calcutta, March 12-14, 1993.*

Yang, P.C. and Ho, F.W. (1980). Effect of embark and ethrel on sugarcane quality, yield and ratoon growth. *Proc. ISSCT*:**17**: 711-724.

Yang, S.J. (1977). Soil physical properties and growth of ratoon cane as influenced by mechanical harvesting. *Proc. ISSCT.* **16**: 835-847.

Index

T

U

V

W

Y